Principles of Insect Parasitism Analyzed From New Perspectives

Practical Implications for Regulating Insect Populations by Biological Means

E.F. Knipling

United States Department of Agriculture

Agricultural Research Service

Agriculture Handbook Number 693

Abstract

E.F. Knipling. 1992. Principles of Insect Parasitism Analyzed From New Perspectives: Practical Implications for Regulating Insect Populations by Biological Means. U.S. Department of Agriculture, Agriculture Handbook No. 693, 337 pp.

This publication presents the results of a theoretical study on the roles that parasitic insects can and cannot play against populations of their insect-pest hosts under natural conditions. More importantly, the feasibility of managing major insect-pest problems through highly augmentative releases of parasites in the host ecosystems is critically examined by theoretical deductive procedures.

Keywords: augmentation release, beneficial insect, biological agent, biological control, entomophagous insect, host insect, insect, insect management, insect pest, insect suppression, integrated pest management, parasitic insect, pest control, pest management, predator insect, prey insect.

Copies of this publication can be purchased from the Superintendent of Documents, U.S. Government Printing Office, Washington, DC 20402. See order form in back of book.

ARS has no additional copies for free distribution.

April 1992

Foreword

The improved use of parasitoids as biological control agents is at the forefront of our quest for more effective, lasting, and environmentally safe technology for pest management. Various avenues are available for utilizing parasitoids, including the importation of new species, mass-propagation and release of parasitoids, and habitat management techniques to increase the abundance and performance of feral and released individuals. Our limited understanding of parasitoid/host interactions and other factors affecting their efficacy is a central barrier to consistently effective use of parasitoids with all of these approaches. The author herein presents an analysis that is of major value in our endeavor to understand host/parasitoid interactions.

Dr. Knipling has for many years advocated the need to seek environmentally compatible and lasting solutions to our major pest problems versus the more conventional, reactive measures and has continually stressed the importance of approaching these pest problems from a total-population perspective. He further has envisioned parasitoids as potentially invaluable weapons in our pest management arsenal and sought ways to develop that potential. This treatise is the fruit from his years of diligent pursuit toward a fundamental understanding of parasitoid/host interactions at the total-population level, and it presents the results that could be derived by the mass-release of parasitoids over a major area of the pest ecosystem.

He has assembled the key information on the variety of factors that are known to affect parasitoid-host coexistence patterns and used it in a thorough analysis of the population dynamics of several parasitoid-host examples, each species having certain unique characteristics. The results derived from these analyses and especially the conclusions as to certain common principles governing parasitoid populations and their efficacy offer a truly new and important perspective. The author's highly significant central conclusion is that the rate of parasitism (at the total-population level) is primarily a function of the ratio of parasitoid numbers to host numbers—with host density, of itself, being of little consequence. A major supporting premise for this conclusion is reported research which demonstrates that parasitoids have highly sophisticated host-detection mechanisms and, thereby, waste very little time searching in spaces other than the immediate sites of host infestation.

It is important to emphasize, as does the author, that because of limited information many of the parameters used in the analyses were derived by deductive reasoning and may be of limited precision. However, the author has taken great care to reduce errors in his assumptions by placing them through rigorous validation trials with numerous models.

As with any such conceptual work, various aspects of the conclusions are certain to be amended and modified with new data and further analysis. In fact, some important sources of variation which have an important bearing on the subject, such as recent findings on learning by parasitoids, are not analysed. These factors, which argue for more dynamics in the models, should be kept in mind as we continue to explore the subject. These considerations do not, however, negate the primary value of the work, which is in the central conclusions; the new perspective for appraising parasitoid/host interactions; and the practical implications for augmentative use of parasitoids. The findings show that augmentation on a total-population scale can be an extremely powerful tool for pest suppression. In fact, when releases raise the overall population parasitization much over 50 percent, a very strong "self-increasing" effect can be set in motion.

Some colleagues may feel that the appraisals and conclusions are oversimplified and too generalized. Such impressions should not, in my opinion, be drawn because of the simplicity of the models. The relatively few parameters used in developing the models are based on a judicious selection of the factors pertinent to the questions addressed. My experience in following Dr. Knipling's various works is that he has a tremendous gift for sorting through complex matters and bringing the essence to focus. I believe he has again achieved that end with this major contribution to the science of biological control.

W. Joe Lewis
Research Entomologist
USDA-ARS
Insect Biology & Population Management
Research Laboratory
Tifton, Georgia 31793

Preface

For many years I have encouraged research and development on techniques and strategies for insect-pest control that in theory offer possibilities of being more effective, more economical, and less environmentally hazardous than those in current use.

The investigation herein described was focused on appraising the role that certain parasites (parasitoids) play as natural control agents, and the role they might play for regulating populations of their pest hosts if the number of parasites in the host ecosystem were sufficiently increased by artificial means. The appraisals were made in the light of (1) important new basic information that insect behaviorists have obtained during recent years on the highly efficient mechanisms that biological agents possess for finding their hosts or prey; (2) reports, by a number of investigators, that for many parasite species there is no evidence of a positive correlation between host density and the proportion of the host population parasitized by the coexisting parasite population; and (3) new, more general information on the biology and behavior of numerous parasite species.

The appraisals were made by deductive procedures, and the results are presented and discussed in detail in this publication. While theoretical, the results are considered to be entirely consistent with the principles of insect parasitism, the factors that influence the dynamics of insect populations, and the basic principles of insect population suppression.

Many complex biological actions and interactions are involved in natural parasitism processes. However, pest hosts and their associated parasites have evolved biological and behavioral characteristics which virtually ensure that the relative numbers of hosts and parasites in the ecosystem they inhabit and the rates of parasitism that occur will remain within rather narrow limits. This is to be expected even though the coexisting populations may fluctuate by up to 100-fold during a single season or over a period of several years for cyclic species. The rather close numerical relationships assumed are entirely consistent with nature's balancing mechanisms, which permit closely associated organisms to coexist in reasonable harmony.

Thus in natural populations, the ratios of parasites to their coexisting hosts are not high enough to result in dependable control of many pest hosts. Without question, however, technology can be developed to efficiently mass-rear parasites of many species so that they may be released at strategic times in host ecosystems in numbers that would result in parasite-to-host ratios high enough to allow effective control of the host populations. To know how many parasites of a given species to produce and release for achieving various rates of parasitism, it would be desirable to have good information not only on the actual and relative numbers of the parasites and hosts coexisting in natural ecosystems but also on the relationship of these numbers to available data on rates of parasitism. For every parasite and its host, elaborate and costly research programs would be required to obtain reliable information on these numbers under the variable conditions existing in natural populations. In this investigation, therefore, theoretical deductive procedures were used to make realistic estimates of the normal numerical relationships between the different parasite-host associations examined. Extensive use was made of rather simple population models, which were developed for each association. To develop the models, every aspect of parasitism and insect population dynamics had to be critically analyzed. Many estimates and assumptions had to be made, based on new theories or the limited information available. The models presented show the numerical relationships between the pest host and parasite populations, the growth rates of these populations, and the parasitism trends over successive generations in natural ecosystems. The basic models were then used to estimate how the parasitism trends and the dynamics of the pest hosts would be affected when the normal numerical relationship throughout a pest ecosystem is greatly increased by artificial means in favor of the parasite population. I doubt whether most of the conclusions in this investigation could have been reached without the aid of the models.

Despite the many unknowns and complex biological actions involved, the theoretical investigation proved to be very fruitful. As discussed in detail in this publication, the parasite release technique has certain unique suppressive characteristics not possessed by any other methods of insect suppression. As we will see, these characteristics have profound practical implications for managing total pest populations when the technique is employed alone or when integrated with other methods of suppression.

Many readers of this publication are likely to question the validity of some of the conclusions reached. They might particularly question the conclusion that parasites have the inherent ability to find hosts readily even when the host population is abnormally low.

They might well ask, Why isn't this ability more evident both from the vast amount of natural parasitism data that are available and from the results of the augmentation experiments that have been conducted in the past?

Most certainly I raised that question—repeatedly—as I analyzed the proposed coexistence patterns of the parasite-host associations included in the study. I looked for flaws in the premises on which the suppression models are based. But when I considered that we can by artificial means create parasite-to-host ratios that exceed ratios in natural populations by as much as 10- or even 100-fold, I gained a great deal of confidence in the theoretical results. As will also be discussed in various chapters that follow, there are a number of sound reasons why the full potential of the parasite release method as a means of regulating total pest populations has remained largely obscured for such a long time.

Very difficult and costly research will have to be undertaken to confirm the results and conclusions of this investigation. Without doubt, the development of the parasite release technique as one of the primary methods for managing insect pest populations will be a challenge for pest management scientists every step of the way. However, given the resources, there should be no other barriers to eventual success in developing this ecologically acceptable way to regulate populations of a wide range of the world's most damaging and formidable insect pests. Current methods for controlling many of these pests are based on using nonselective insecticides and therefore cause extensive ecological disruptions. For many parasite-host associations, I estimate that the costs involved to maintain the pest populations below economically damaging levels would be only a fraction of the losses the pests cause to agriculture under current management procedures. The parasite release technique offers not only opportunities to avoid environmental hazards but also the realization of large savings to agriculture. Nature has made available hundreds of parasite species that would be suitable candidates for exploiting this biologically sound and environmentally safe means of regulating insect pest populations.

Acknowledgements

I am indebted to many individuals for information and assistance in the conduct of these investigations. I received much valuable information on the biology and behavior of the various parasites and hosts from my associates with the Agricultural Research Service, U.S. Department of Agriculture. They include W. Joe Lewis, Harry R. Gross, William C. Nettles, Edgar G. King, James E. Gilmore, Tim T. Wong, Sess D. Hensley, William E. Guthrie, Ralph E. Webb, Alton N. Sparks, John J. Drea, Jack R. Coulson, Thomas R. Ashley, and Keith R. Hopper. I also appreciate the information received from Thomas M. Odell, Forest Service, U.S. Department of Agriculture. In addition, I received valuable information from colleagues with State institutions, including H.C. Chiang and Richard L. Jones, University of Minnesota; R.M. Baranowski, University of Florida; and James R. Cate, Texas A&M University. Finally, I am deeply indebted to Dee Haley and Frances James for typing and helping to prepare the manuscript, and to Alice Kunishi for editorial assistance.

Contents

Principles Of Insect Parasitism Analyzed From New Perspectives

Practical Implications for Regulating Insect Populations by Biological Means

E.F. Knipling

1 Introduction

Thousands of parasitic and predator insects have evolved in nature and help to prevent insect pest populations from expanding without limit (see fig. 1). But at the same time, the populations of these beneficial insects are also kept within bounds by nature's unique and stringent balancing mechanisms. Individually and collectively, therefore, beneficial insects are unable to reduce the populations of many insect pests below levels that are economically damaging to agricultural interests or that seriously threaten the health and comfort of humans and animals. Also, human activities that result in the creation of artificial environments often tilt the balance in favor of the pests or potential pests and further limit the effectiveness of their coexisting parasites and predators.

The concept of controlling insect pests by increasing the number of parasites or predators in natural environments by artificial means has long been considered promising (Biliotti 1977). Indeed, this augmentation procedure is employed with varying degrees of success in many parts of the world (Dysart 1973, Beglyarov and Smetnik 1977, Biliotti 1977, Huffaker 1977, Ridgway et al. 1977, Coulson et al. 1979, Hassan 1988). However, with the exception of *Trichogramma* spp., very few of the many species that would seem to be excellent candidates for controlling pest species by this procedure have been investigated. When the augmentation or inundative procedure is employed, the parasite releases are generally made in limited pest habitats. The objective is somewhat similar to that of the long-practiced procedure of applying insecticides only where and when they are needed to alleviate damage.

See p. 337 for biographical sketch of author.

Biological agents may be augmented in natural environments in a number of ways (Huffaker et al. 1977). The investigation reported herein appraises the feasibility of regulating total populations of agriculturally important pest insects by the release of select, mass-reared parasites in the pest-host ecosystems. Therefore, we will be concerned with augmentation in large areas of pest habitats—preferably, isolated or entire pest ecosystems.

It is my view that if major advances are to be made in coping with certain important insect pest problems, the tactics and strategies for managing the insects must change. They must change from the largely limited, reactive measures used heretofore to rigid, preventive measures that are target-pest specific. That is, *suppressive measures must be applied against total pest populations in an organized and coordinated manner at strategic times before—and not after—the populations have reach damaging numbers.* Parasite augmentation would seem to be highly desirable as a preventive measure, because if the parasites released are host specific, they will pose no danger to humans, beneficial organisms, or the environment. To my knowledge, no *experiments have ever been undertaken to test the concept of total-pest-population management by the release of large numbers of parasites throughout pest-host ecosystems.* Thus, the role that such parasite releases can play in total pest-population management procedures is unknown.

Most parasite release experiments conducted in the past have been undertaken in small unisolated areas. Yet, most pest insects, as well as the parasites or predators that depend on the pest insects for survival, are highly mobile. The released organisms and their progeny may be capable of dispersing for miles during the course of a host generation or during a crop-growing season. Is it logical to expect definitive results from augmentation experiments when releases of highly mobile parasites are limited to individual farms or even to sizable communities? I believe the answer to the question is self-evident. For valid results, the parasites must be released in adequate numbers at the proper time during the seasonal or periodic life cycles of the pest hosts and in areas large enough to greatly minimize the effect of unilateral movement of the released insects and their progeny out of the release area. Of comparable importance is the possible excessive movement of the pests into areas where protection is needed. Therefore, depending on the dispersal behavior of a parasitic insect and its host, valid augmentation experiments may have to be conducted on a regional scale and well in advance of the appearance of the pest in damaging numbers.

Figure 1. *Microplitis croceipes.* This biological agent parasitizes larvae of *Heliothis zea* and *H. virescens*, two of agriculture's most damaging insect pests. *M. croceipes* is but one of hundreds of parasite species that depend on insect pests for their survival. Without natural biological agents, insect-pest control would be much more difficult and costly than it now is. However, because of nature's balancing mechanisms, many biological agents cannot reach population levels high enough to effectively control their hosts or prey.

The theoretical investigation described in this publication focuses on several parasite species and their associated hosts. For each parasite-host association, estimates are made of (1) the numerical relationship between the natural populations of the parasite and host in a typical pest-host ecosystem and (2) the numbers of parasites that would have to be reared and then released in the pest-host ecosystem if populations of the pest host are to be kept below levels that cause economic damage to crops. The results of the investigation indicate that it should eventually be technically and economically feasible to rear and release enough parasites of a wide range of species to prevent such damage.

Most agricultural insect pests have coevolved with parasite species that depend on the insect pests for reproduction and survival. Therefore, the parasite augmentation approach to pest population management offers the possibility of coping with many major insect-pest problems not only more effectively and at less cost than current management practices, but also without health risks and without harm to the many beneficial organisms in our agricultural ecosystems. (USDA photograph.)

Unfortunately, however, research resources are usually inadequate to undertake augmentation experiments in the manner and scale necessary to determine the potential of parasite releases for regulating pest populations. Before appropriate experiments can be undertaken, it may be necessary to develop mass-rearing methods for the parasites to be evaluated. Understandably, administrators as well as scientists are generally reluctant to commit substantial financial resources for specific research purposes unless there is tangible evidence that the proposed research will be productive. But such evidence cannot be obtained without appropriate field experimentation. Thus, the potential of literally hundreds of parasites for managing pest-host populations remains unknown.

As an alternative, or preliminary, to conducting release experiments, I used theoretical deductive procedures to appraise the potential of select parasite species for regulating populations of some of our major insect pests. If certain parasites can be shown to offer excellent prospects for effectively controlling major pests, the research community and society in general may be more willing to accept the costs and risks involved in developing and testing the augmentation approach. Such appraisals have been made in the past (Knipling 1979). But since then, research scientists have gained a much better understanding of the biology and behavior of insect parasites, as evidenced by the series of articles in books edited by Ridgway and Vinson (1977), Nordlund et al. (1981), Waage and Greathead (1986), and others. Researchers have also obtained more information on the biology, ecology, and dynamics of many insect pests and their associated parasites. This new knowledge points out the need to reconsider the factors that govern parasite-host relationships and interactions in natural populations, and their effect on the role that parasites can play in pest-host control when the number of parasites is greatly increased by artificial means.

As the theoretical investigation progressed, I reached the conclusion that environmental constraints imposed by nature's balancing mechanisms rigidly limit the actual and relative numbers of parasites that can coexist with the host populations and that the constraints therefore contribute in large measure toward preventing self-perpetuating, or natural, populations of many parasite species from achieving the rates of parasitism necessary for dependable control. This conclusion implies that the rate of parasitism achieved by a given species is not a reflection of its host-seeking ability but, rather, of the relative numbers of the parasites and hosts that normally coexist when the two organisms have reached stability with the environment and with each other. I believe that biologists can accept with certainty the hypothesis that all host-dependent species search diligently and effectively, regardless of the density of the host population. Otherwise, they will not have survived for many thousands of years, during which catastrophic events doubtless reduced their populations to abnormally low levels from time to time. The hypothesis leads me to believe that the numerical barrier that nature imposes can be effectively breached by greatly increasing the parasite populations at a strategic time in the host cycle. In theory, we can show that if the ratio of parasites to hosts achieved by augmentation is high enough initially, the regulating forces of nature cannot come into play in time and with sufficient intensity to prevent dramatic declines in the host populations before the normal numerical relationships can be restored. The above, in brief, is the rationale on which this study is based.

Key data desired for my study were the absolute numbers of parasites and hosts that coexist in nature. But because such data are usually unavailable, population models of parasite-host associations were developed by a two-step procedure: (1) the numbers of hosts present in a given area during their annual or periodic cycles were estimated in the light of available data on pest abundance and general knowledge of the dynamics of the major pest insects and (2) the estimated numbers were, in turn, used to estimate the likely numbers of known coexisting parasite species, based on the factors influencing natural parasitism that will be discussed. In the course of developing these models, I gained a much broader perspective of parasite-host relationships and interactions. My views of the roles that host-specific parasites can and cannot play in regulating their host populations and of the factors that determine the role they do play underwent considerable modification from those I held previously (Knipling 1977, 1979). I developed literally hundreds of preliminary, stochastic population models based on different values for the parameters considered relevant to the various parasite-host associations. Then, the models that seemed the most realistic in terms of relative numbers, population trends, and normal rates of parasitism were selected. Using these models, I extrapolated the numbers of parasites that would have to be released in the pest-host ecosystems to raise the rates of parasitism to various levels.

The extrapolated values have relevance to the need for efficient insect-rearing technology. The feasibility of making practical use of a parasite for augmentation purposes will depend on the demographics of the host and parasite populations, the number of parasites required, and the costs that will be involved in rearing and distributing the parasites. In view of the advances that have already been made in insect-rearing technology (King and Leppla 1984), however, there should be few technical barriers to the rearing of most candidate parasite species in unlimited numbers and at reasonable costs. While in vitro mass-rearing procedures would no doubt greatly advance the augmentation technique, the estimated efficiency for select species suggests that it should be possible to economically mass-rear many parasite species on the hosts.

By estimating the numbers of parasites required to achieve high rates of parasitism and relating the estimates to probable costs of mass-rearing various parasite species, I reached the conclusion that chances are excellent that populations of a wide range of agricultural pest insects can be effectively regulated at a mere fraction of the monetary losses they cause under current management practices.

The augmentation method is one of a number of ecologically sound methods of insect-pest management that scientists are making concerted efforts to develop. These efforts were prompted several decades ago by the realization that despite the remarkable advances made in insect control by the many highly effective insecticides that first came into being during the mid-1940's, the insecticides commonly used create many complex and serious environmental problems. Insecticide resistance continues to cause additional problems. Also, the dynamics of most insect pests are clearly such that an attack on only a limited proportion of the total pest population, no matter how intensive or for how long, will not resolve the pest problems. Indeed, if broad-spectrum insecticides are used, such tactics may intensify the threat of the target pest, as well as associated potential pests, because of the simultaneous destruction of many beneficial biological agents.

In the search for alternative methods of insect control, scientists have made outstanding progress in a number of areas. They have learned much about the methods insects use for sexual communication. This knowledge has lead to the identification and synthesis of sex pheromones of many important insect pests. These attractants can be used in the development of highly sensitive methods to detect pest insects. If appropriately used, they also offer prospects for the development of methods to control pests. Insect geneticists and entomologists have discovered ways that might be used to produce genetically altered insects that would be much more effective than sterile insects in reducing reproduction. Their discoveries increase the opportunity for making greater use of autocidal control procedures. Progress has also been made on conventional methods of control, including the use of more selective insecticides, pest-resistant crop varieties, and more effective insect pathogens. New biological agents introduced for permanent establishment help alleviate certain insect problems. There is increasing interest in controlling pests by the augmentation procedure.

In general, however, the insect pests that were of major importance to agriculture 35–40 years ago are still of major concern today. In effect, much of the new information and the advances in insect suppression technology have not yet been put into practice. The effectiveness of certain control measures is strongly influenced by the density of the target pest population (Knipling 1979). Decreasing effectiveness with increasing pest density is a characteristic of autocidal techniques; techniques using sex attractants; and, as will be discussed in detail in this publication, the parasite augmentation technique. Also, these techniques are slow acting as compared with insecticides. Therefore, unlike insecticides these alternative techniques are not likely to be effective or practical after the pest populations have reached economically damaging levels.

For these reasons, it is my conviction that much of the insect suppression technology we have or that might be developed in the future will not play a prominent role in resolving major insect-pest problems unless it is applied against total populations as preventive measures and in areas large enough to greatly minimize the insect movement factor. Otherwise, growers will have no recourse but to rely on fast-acting remedies where and when the need arises; and such remedies usually require the use of ecologically disruptive chemical insecticides.

The results of the theoretical investigation to be presented in this publication give reason to believe that the parasite augmentation technique offers outstanding possibilities as a pest-management technique. Theoretically, as will be shown, the progeny produced by the released parasites will generally be much more important than the original releases in keeping the pest-host populations below economically damaging levels. Indeed, continuing control of a pest for several pest cycles by the self-perpetuating populations resulting from the initial releases is a unique characteristic of this biological method of pest management. The suppressive action will continue until natural regulating forces again restore a normal balance between the parasite and host populations. It would be difficult to envision a more biologically sound approach to insect-pest management than to employ target-pest-specific biological organisms that have evolved naturally. Their use will protect the quality of the environment and permit maximum benefit from the other natural biological organisms present in pest ecosystems. This has been a long-sought goal.

The theoretical results are so impressive that their validity will certainly be questioned. This is to be expected because the extrapolations go far beyond any available data to support them. However, the numerical relationships brought about by augmentation to the degree proposed cannot occur naturally. Therefore, conclusions based on the limitations of natural, self-perpetuating parasite populations for regulating pest populations are not valid. Nevertheless, since the technique has never been tested in the manner proposed, we do not and will not know how realistic the results are until the augmentation technique is properly tested. However, if actual results merely approach the theoretical, the parasite augmentation technique used in the manner proposed should offer almost unlimited potential for coping with many of the world's major pests effectively and practically, especially when integrated with autocidal and other compatible techniques.

In the various chapters that follow, some of the more important factors that govern parasite-host relationships will be critically analyzed; how the key factors act in unison or interact to allow the members of a parasite-host association to coexist will be described; models will be presented that depict the estimated numerical relation and population dynamics of several major pests and their associated parasites as they coexist in nature; and corresponding, suppression models will be presented that depict the estimated effects of greatly increasing parasite numbers throughout their pest host ecosystems.

The parasite associations evaluated in this investigation are listed at the end of this chapter. The biology and behavior of the parasites are diverse, as are those of the pest hosts. If the augmentation technique proves to be effective and practical for controlling most of the pests listed, it should also be applicable to many other pest species.

The reader is urged to read chapters 2 and 3 before reading the chapters of special interest.

Pest host	Associate parasite	Parasite characteristics
Sugarcane borer (*Diatraea saccharalis* (Fab.))	*Lixophaga diatraeae* (Townsend)	Solitary larval parasite; usually parasitizes 3d- to 5th-instar larvae.
European corn borer (*Ostrinia nubilalis* (Hübner))	*Microcentrus grandii* (Goidenich)	Polyembrionic larval parasite (ca. 20 parasites develop per host); parasitizes 3d- to 5th- instar larvae.
Mediterranean fruit fly (*Ceratitis capitata* (Wied.))	*Biosteres tryoni* (Cameron)	Solitary larval parasite; oviposits in 3d-instar larvae.
Oriental fruit fly (*Dacus dorsalis* (Hendel))	*Biosteres arisanus* (Sonan) = *Opius oophilus* (Fullaway)	Solitary egg-larval parasite; oviposits in eggs.
	B. longicaudatus (Ashmead)	Solitary larval-pupal parasite; oviposits in 3rd-instar larvae.
	B. vandenboschi (Fullaway)	Solitary larval parasite; parasitizes 1st- and 2d-instar larvae.
Caribbean fruit fly (*Anastrepha suspensa* (Loew))	*B. longicaudatus*	Solitary larval-pupal parasite; oviposits in large larvae.
Fall armyworm (*Spodoptera frugiperda* (J.E. Smith))	*Chelonus insularius* (Cresson)	Solitary egg-larval parasite; oviposits in eggs.
Gypsy moth (*Lymantria dispar L.*)	*Blepharipa pratensis* (Meigen)	Solitary larval-pupal parasite; oviposits on leaves of host plant.
Heliothis zea[1] (Boddie) and *H. virescens* (Fab.)	*Microplitis croceipes* (Cressan)	Solitary larval parasite; oviposits in 2d- to 4th-instar larvae.
	Archytas marmoratus (Townsend)	Solitary larval-pupal parasite; larviposits on leaves of host plant.
	Eucelatoria bryani (Sabrosky)	Exhibits multiparasitism behavior (ca. 2.5 larvae per host); larviposits in large host larvae.
Boll weevil (*Anthonomus grandis* (Boheman))	*Bracon mellitor* (Say)	Solitary larval parasite; oviposits in 3d-instar larvae
	Catolaccus (= *Heterolaccus*) *grandi* (Burks)	Solitary larval parasite; oviposits in 3d-instar larvae.

[1]The nomenclature *Heliothis zea* is used throughout this publication. However, the nomenclature accepted by most current taxonomists is *Helicoverpa zea*.

2 Factors Influencing the Behavior and Efficiency of Host-Specific Parasites

The relationships that have evolved between parasites and their host populations in the ecosystems they cohabit are variable and complex. This is evident from the information on behavior of parasites contained in publications such as those of Ridgway and Vinson (1977), Nordlund et al. (1981), and Waage and Greathead (1986). Every parasite-host association has evolved coexistence patterns that tend to optimize the reproductive success of the parasite without unduly risking the welfare of the host. Excessive parasitism to the extent of jeopardizing the welfare of the host would not be advantageous to the parasite. It is generally accepted that factors associated with nature's balancing scheme tend to maintain the population of each organism at what is called a steady density or characteristic density (Huffaker et al. 1977). As it applies to this discussion, the concept of steady density means that parasites and their hosts establish population densities that permit coexistence in reasonable harmony.

Despite the many successes of introduced insects in controlling insect pests (Clausen 1956, 1978; De Bach 1964; and Sailer 1972), the natural rates of parasitism of most of the introductions—and, indeed, of many endemic parasites—are not high enough for adequate control. This we know from decades of observing a wide range of important insect pests. No doubt, the steady densities of the parasites and hosts limit the rates of parasitism that can be achieved. How can the effectiveness of parasites be increased to the extent that they will adequately control insect pest populations? Clues to the answer might be obtained by examining the factors that influence or determine the numerical relation between parasite and host populations and the rates of parasitism. In the following sections, the various factors that are known or thought to be involved are analyzed separately. Some of the factors interact and thereby increase greatly in importance.

Host Density

No other aspect of parasite-host relationships has received more consideration by biologists than the influence of host density on the behavior and efficiency of the associated parasites. The rates of parasitism that occur at various host densities have been the subject of special interest. The prevailing concept for many years seemed to be that the host-finding efficiency and the rate of parasitism caused by a coexisting parasite population tend to increase as the host density increases. There is increasing evidence, however, that for many, if not most, parasite species, the percentage of the host population parasitized is not positively correlated with the host density. A few publications will be cited as evidence. Miller (1963) reported that changes in the larval density of the spruce budworm, *Choristoneura fumiferana* (Clem.), had no consistent effect on the rates at which the larvae were parasitized by two common parasites, *Apanteles fumiferaneae* (Vier.) and *Glypta fumiferaneae* (Vier.). Torgensen et al. (1984) studied the rates at which several species parasitized the larval and pupal stages of the western spruce budworm, *C. occidentalis* (Freeman), when the host populations were sparse and stable, in an increasing outbreak mode, and in a stable outbreak mode. They found no positive correlation between rate of parasitism and host density but did find a tendency for the rate of parasitism to increase during the period of host decline following an outbreak. As will be discussed elsewhere, such an increase in parasitism cannot be attributed to a positive response to host density.

The rate of parasitism of the oriental fruit fly, *Dacus dorsalis* (Hendel), by *Biosteres arisanus* (= *Opius oophilus*) has been under observation for many years in Hawaii. This and other parasite species that begin their development in host eggs and complete their development when the host is in the larval stage are referred to as "egg-larval parasites." Bess et al. (1963), Newell and Haramoto (1968), and Haramoto and Bess (1970) found no correlation between the average rate of parasitism and host density, fruit density, or season of the year. Although host and fruit densities fluctuate considerably from season to season and year to year, parasitism has remained near 60 percent for many years. Wong and Ramadan (1987) recorded the rates at which oriental fruit fly and Mediterranean fruit fly larvae in Hawaii were parasitized by several species during the months of January to July, when the host densities vary considerably. No consistent difference in the average rates of parasitism seemed evident for the period of record.

Gargiullo and Berisford (1981) investigated the rates at which eight species of Hymenoptera parasitized larvae of the southern pine beetle, *Dendroctonus frontalis* (Zimmerman). For all species except *Spathius pallidus* (Ashmead), changes in host denisty either had no effect on or were inversely related to the rate of parasitism. The rate of increase of parasite density never exceeded that of its host's density.

Also difficult to find is clear evidence for a positive correlation between the rate of parasitism due to a complex of species and the density of their host population. Fuester et al. (1983) could find no such evidence in their study of several species of parasites and their larval gypsy moth hosts in Burgenland, Austria, and Wurtsburg, Germany. However, as seems characteristic of many complexes of parasite species, the rate of parasitism seemed to increase during the declining phase of the host population. Puterka et al. (1985) conducted a broad study of parasites of *Heliothis* spp. on a variety of crops in Texas during the crop-growing season. They detected no significant correlation between larval densities and rates of parasitism. In Arkansas, Burleigh and Farmer (1978) found the parasitism rate of *Heliothis* larvae to be fairly constant during July and early August but to decline during late August and September. The use of insecticides may have contributed to this late-season decline. Browning and Oatman (1984) reported that the parasitization rate by the parasite complex of the cabbage looper, *Trichoplusia ni*, was quite stable during the season. They observed only minor changes in species composition with changes in crop type and season.

Dempster (1983) studied the life tables of 24 Lepidoptera. In only three cases did he consider natural enemies to be acting in a density-dependent manner. Morrison and Strong (1980) discussed the possible relationship between spatial variation in host density and intensity of parasitism. They consider a positive correlation between these parameters to be uncommon in nature; a negative or lack of correlation is reported more frequently. Other examples will be given which indicate little or no positive correlation between host density and rate of parasitism when different parasite-host associations are discussed.

This very brief review suggests that as regards specific biological organisms, biologists need to reevaluate the influence of host density on the rate of parasitism. It is my opinion that many, if not most, host-dependent species have evolved capabilities that virtually eliminate host density as a factor governing the rates of parasitism, at least so long as the host populations are within normal levels. It is recognized however that to a considerable extent, host or prey densities must be positively correlated with the proportion of the pest population killed by the total complex of entomophagous insects. Otherwise, in favorable environments, pest densities would tend to increase to infinite levels.

For the parasites evaluated, reasons will be offered as to why host density is not a major factor when the coexistence patterns of these parasites and their associated hosts are discussed. It is noted here, however, that a general lack of positive correlation between rates of parasitism and normal host densities would have very important practical implications. It would mean that the efficiency of a given number of parasites will be inversely proportional to the host density. Such an inverse relation is characteristic of autocidal techniques and techniques using sex pheromones. If it proves to be characteristic of parasite releases, it will be contrary to the views I previously held (Knipling 1979). The influence of host density will be discussed in greater detail in various other chapters of this publication.

Parasite Density

A basic hypothesis adopted for this investigation is that the females in a parasite population search for hosts individually and independent of other conspecifics in the population. Therefore, for example, if 1 female under given conditions is capable of parasitizing 50 hosts during its lifetime, 10 will parasitize 500; 1,000 will parasitize 50,000; and so forth. This would seem to be a logical hypothesis. However, the proportion of the host population that will be parasitized and the proportion that will escape parasitization by chance will depend on the total number of parasite-host encounters in relation to the total number of hosts present. If the females search independently, the higher the parasite population in relation to the host population, the greater will be the ratio of encounters to hosts present; and the greater this ratio, the greater will be the probability that a previously parasitized host will be encountered by the searching parasites. If a host previously parasitized by a conspecific is found, the parasite may reject the host if it has discrimination behavior. Otherwise, the parasitized host may be reparasitized.

Fiske (1910) discussed the influence of superparasitism on the efficiency of parasites. He pointed out, for example, that if 50 percent of a host population has already been parasitized by conspecifics, the probability will be 50 percent that a searching parasite will encounter a previously parasitized host. As a general rule, a host that is parasitized by a solitary species—namely, a species that deposits one egg or larva per host per attack—will permit the development and survival of only one parasite progeny. Usually, therefore, superparasitism by solitary species will be redundant and will not contribute to reproductive success. In fact, as will be discussed later, excessive reparasitization by some species may reduce the probability of survival of the immature parasites. The ability or inability of a parasite species to avoid superparasitizing its hosts is therefore a significant factor influencing the efficiency of the species.

Unless otherwise stated, I make the assumption that the probability of parasites to encounter hosts previously parasitized by conspecifics represents a complete loss in host-parasitization efficiency. While discrimination behavior, which will be considered later, may minimize the influence of this factor, the time and energy required to locate previously parasitized hosts will still be lost.

Since the probability of encountering previously parasitized hosts depends on the rate of parasitism that has already occurred, intraspecific competition must be recognized as one of the important natural regulating factors that limit the rate of parasitism and the reproductive success of self-perpetuating parasite populations. Also, as we will see when estimates are made of the influence of highly augmented populations, the probability of encountering previously parasitized hosts will theoretically have a profound adverse effect on the rate of successful parasitism that is likely to be achieved.

In view of the significance of intraspecific competition on the reproductive success of parasite populations, table 1 is included to show how the ratio of total parasite-host encounters to total hosts present relates to the proportion of the host population that will be parasitized one or more times and to the proportion that will escape parasitization by chance. The data apply to coexisting parasite-host populations during a period corresponding to one generation. Frequent reference will be made to the table in discussions on the influence of augmented parasite populations on host populations. A few examples will be given to indicate the importance of the intraspecific competition factor. If the ratio of the number of parasite-host encounters to the number of hosts present is 1:10, actual parasitism will be 9.5 percent (including superparasitism), and 90.5 percent of the hosts will escape parasitism. When the ratio is as high as 1:1, approximately 63.4 percent will be parasitized one or more times, and 36.6 percent will escape parasitism by chance. A ratio of 2:1 will theoretically result in 86.6 percent parasitism, and 13.4 percent will escape parasitism.

The probability data clearly indicate the increasing loss in efficiency of a parasite population as the ratio of parasite-host encounters to

Table 1
Percentage of a host population that will be encountered 1 or more times and, presumably, parasitized by a parasite population, and the percentage that will escape parasitism by chance, depending on the ratio of the total number of times hosts are encountered to the total number of hosts present in a pest-host ecosystem[1,2]

Ratio of parasite-host encounters to hosts present	Percentage of hosts parasitized (unless parasite discriminates)	Percentage of hosts that escape parasitization by chance
1:100	1	99
1:50	2-	98+
1:33.3	3-	97+
1:25	4-	96+
1:20	4.9	95.1
1:10	9.5	90.5
1:6.7	13.9	86.1
1:5	18.1	81.9
1:4	22.1	77.9
1:3.3	25.9	74.1
1:2.9	29.5	70.5
1:2.5	33.0	67.0
1:2.2	36.2	63.8
1:2	39.3	60.7
1:1.8	42.3	57.7
1:1.5	47.8	52.2
1:1.4	50.3	49.7
1:1.3	52.8	47.2
1:1.25	55.1	44.9
1:1.18	57.3	42.7
1:1.1	59.3	40.7
1:1.05	61.3	38.7
1:1	63.4	36.6
1.1:1	66.7	33.3
1.2:1	69.9	30.1
1.3:1	72.7	27.3
1.4:1	75.3	24.7
1.5:1	77.7	22.3
1.6:1	79.8	20.2
1.7:1	81.7	18.3
1.8:1	83.5	16.5
1.9:1	85.0	15.0
2:1	86.6	13.4
2.5:1	90.8	9.2
3:1	95.1	4.9
4:1	98.2	1.8
5:1	99.3	0.7

[1]Values for ratios not shown can be estimated by interpolation.
[2]I am indebted to J.U. McGuire (U.S. Department of Agriculture, retired) and to R.R. Knipling for assistance in developing the following formulas to derive the percentage of hosts parasitized 1 or more times and the percentage that escape parasitization by chance:

Percentage parasitized = $100(1 - e^{-x})$
Percentage not parasitized = $100(e^{-x})$
When

x = ratio of total parasite-host encounters to total hosts present
e = natural logarithm base (2.718)

The formula for deriving the ratio x is:

$x = \frac{rt}{r_1 t_1}$, when

r = number of female parasites in population

t = average number of hosts (eggs or larvae) encountered by each female parasite during its lifetime

r_1 = number of female hosts in population

t_1 = average number of hosts (eggs or larvae) produced by each female host during its lifetime

Example of calculations:

Assume r = 160; t = 40

$r_1 = 240; t_1 = 50$

$x = \frac{160 \times 40}{240 \times 50} = \frac{6{,}400}{12{,}000} = 0.53$

Percentage parasitized $= 100(1 - e^{-x}) = 100(1 - e^{-0.53})$

$= 100(1 - 0.589) = 100(0.411)$

$= 41.1$

Percentage not parasitized = 100 – 41.1 = 58.9; or,

$100(e^{-x}) = 100(e^{-.53}) = 100(0.589) = 58.9$

hosts present increases. One may consider the effect in another way. After parasitism has reached 10 percent, only 1 of 10 hosts found thereafter would be unsuitable for successful parasitism, a 10-percent loss in efficiency. If, however, parasitism has already reached 90 percent, 9 of 10 hosts found thereafter would be unsuitable for successful parasitism, a 90-percent loss in efficiency. These simple examples indicate how one of nature's regulating forces operates to minimize the probability of excessive parasitism. The intraspecific competition factor alone would seem to make it highly unlikely that a parasite species will be able to achieve the degree of reproductive success necessary to reach and maintain high rates of parasitism. Other factors would also become involved. When parasite populations reach levels high enough to cause high rates of parasitism, they are likely to suffer a higher rate of predation and, hence, a shorter reproductive life. Understandably, therefore, consistently high rates of natural parasitism are rare among most parasite-host complexes.

During many years of coexistence with a host-dependent parasite species, host populations have undoubtedly evolved behavior patterns that tend to ensure relatively low rates of parasitism. However, even under unusual conditions, the laws of chance probably also ensure that the survival of a host will not be seriously jeopardized by a self-perpetuating parasite population. According to table 1, if the total number of parasite-host encounters exceeds the total number of hosts by a ratio of 3:1, 5 percent of the hosts would still escape parasitism by chance. For solitary parasite species, there would be virtually no possibility for the ratio of adult parasites to adult hosts in natural populations to reach levels that could result in 95 percent parasitism. However, as we will discuss later, if enough parasites are released to increase the parasite population by up to 25-fold or more, there may be no way for the host population to prevent near-100-percent parasitism despite the laws of chance and the intraspecific competition factor.

Besides intraspecific competition, a host population in a natural environment faces many other biotic hazards. The rate of parasitism due to all other parasites attacking the host is likely to be greater than the rate of parasitism due to any one species. The interspecific competition factor undoubtedly contributes to a reasonable balance of all species. A seemingly popular concept is that all biological agents that attack a given pest host are beneficial. This is not entirely true. Parasitism that preempts successful parasitism by other parasite species, or predation that destroys immature parasites along with the hosts, will make no significant contribution to natural biological control.

In the absence of such "enemies," a single, well-adapted parasite might, in some cases, provide a higher degree of control than the total complex of biological agents. The addition of adult parasites into pest-host ecosystems would be a means to largely compensate for the adverse influence of interspecific competition.

Kairomones

Recent research on the host-searching behavior of parasites (Nordlund et al. 1981, Waage and Greathead 1986) has contributed to a better understanding of the parasitization process in natural populations. While much of the information on parasite behavior is qualitative, it nevertheless serves as an important basis for estimating the host-finding capability of parasites in quantitative terms. Host finding by parasites is a rather precise process. Vinson (1981) describes the various steps involved in host parasitization. Three rather distinct steps are recognized, namely, host habitat location (Vinson 1981), host location (Weseloh 1981), and host acceptance (Arthur 1981). A basic premise is that a species that is totally dependent on a given host for reproduction is a highly specialized organism having exceptional host-finding capabilities. While several detection mechanisms may be used, the most effective appear to involve chemical cues produced by the host or plant harboring the hosts. These chemical cues are called kairomones.

The chemical nature of some kairomones has been identified (Jones et al. 1971, Lewis and Jones 1971, Greany et al. 1977, Vinson 1977, Jones 1981). Lewis et al. (1975) and Loke and Ashley (1984) determined that kairomone signals intensify host searching and increase the host-finding capability of parasites. The use of kairomones offers the possibility of manipulating natural populations of parasites and predators to achieve more effective insect-pest management, as discussed by Gross (1981) and Lewis (1981). Kairomones may also be useful for manipulating parasites released for control purposes.

The knowledge that parasites possess highly developed host-guidance mechanisms is, in my view, among the most important developments in the field of biological control of insects. These mechanisms help explain the lack of correlation between host densities and rates of parasitism among many parasite-host associations. The chemical cues may enable the females of a host-dependent parasite population to find their normal lifetime quota of hosts even when the host population is abnormally low. The effect of changes in host density on the relative host-finding capability will depend largely on the relative amounts of time the parasites devote to host searching in habitats where hosts are present and where they are absent. As we will consider in greater detail elsewhere, the host-guidance mechanism has very important practical implications in the regulation of insect pest populations by the release of parasites, especially highly mobile parasite species.

Probably few, if any, parasites could survive if they relied on random host finding. Such a reliance would mean that the number of hosts found during a given period of host searching would be proportional to the number of infested and uninfested plants. In efforts to develop models that reflect parasitism trends reported for various parasite-host associations, it soon became apparent that realistic models cannot be developed if it is assumed that the average number of hosts parasitized during the life of the females tends to increase or decrease proportionally to the increase or decrease in host populations. Rates of parasitism would approach zero within a few cycles if host densities exist for several generations at abnormally low levels and would approach 100 percent within a few cycles if the host densities remain abnormally high. Reliance on highly sensitive host-detection signals will, on the other hand, permit parasite and host populations to coexist in reasonable harmony and ensure that enough hosts will be parasitized to maintain viable parasite populations, even though the host densities fall to abnormally low levels from time to time.

Roth et al. (1982) observed the relative amounts of time female *Lixophaga diatraeae* spent on plants harboring larvae of the sugarcane borer and on plants not infested with the larvae. The females spent 7.5 times as much time on infested plants as on uninfested plants. The observations by Loke et al. (1983) are even more impressive. *Cotesia margiventris* devoted approximately 33 times more time searching for hosts on plants harboring fall armyworm larvae than on uninfested plants. W.J. Lewis, U.S. Department of Agriculture, estimates that *Microplitis croceipes* devotes less than one-tenth as much time searching for *Heliothis* larvae on uninfested plants as on infested plants (personal communication).

Such information can be used to make gross estimates of the relative average numbers of hosts individual parasites will be capable of finding at different host densities when all other factors are equal. For example, we will consider the European corn borer (ECB) as a model host, and we will assume that the amounts of time females of a parasite species spend on infested plants and on uninfested plants are in the ratio of 10:1. According to McGuire et al. (1957), the larvae of ECB are distributed largely at random in corn fields. We may assume a corn plant density of 20,000 per acre and two larval host densities—20,000 and 2,000 per acre. To simplify the calculations, we will assume that only one larva will infest a plant. Thus, at the high larval density, all the plants harbor larvae; but at the low larval density, only 2,000 plants per acre harbor larvae, and the ratio of infested plants to uninfested plants would be 1:9. Then, the following formula may be used to calculate the relative host-finding capabilities of the female parasites for both the high- and low-density populations:

$$\text{Host-finding efficiency (percent)} = \frac{\text{Total time spent on infested plants}}{\text{Total time spent on infested and noninfested plants}} \times 100$$

Substituting for the values given, the theoretical results would be 100 percent for the high density and as follows for the low density:

$$\frac{1 \times 10 = 10}{(1 \times 10) + (9 \times 1) = 19} \times 100 = 1{,}000/19 = 53 \text{ percent}$$

Theoretically, therefore, during a given period of host searching, a parasite population under the conditions described would parasitize 53 percent as many hosts at the low density as at the high density despite the 10-fold difference in host densities. This estimate of 53 percent is based on the assumption that the parasites devote equal time to host searching at the two host densities. Probably, however, the parasites could find more than their normal daily or lifetime quota of hosts when all or most of the plants harbor larvae. When the host density and ratio of infested plants to uninfested plants are low, the parasites may merely have to search for a longer period each day or during their lifetime to compensate for fewer hosts found per unit of searching time. In natural environments, other factors besides chemical cues for host detection would also tend to favor host finding when host densities are low. For example, the number of ECB larvae present per acre is likely to be lower by severalfold during the first brood than during the second. However, the size of the corn plants will be considerably smaller during the first brood than during the second. Therefore, the exact location of the host larvae may be detected by the parasite with less searching time during the first brood. If all of these factors are considered, the average number of hosts parasitized per female parasite is likely to be fairly constant though there may be as much as a 10-fold variation in host density.

While the foregoing analyses may have flaws, I believe the basic assumptions are sound. In fact, they are supported by available parasitism data for one of the most common parasites of ECB larvae, *Eriborus tenebrans* (Gravenhorst). Hill et al. (1978) obtained data in Nebraska on the rates of parasitism caused by this species during 1964–71. The ECB densities were also recorded. There is no evidence of a positive correlation between the recorded rates of parasitism and number of ECB larvae per acre. Host densities during 8 years ranged from about 5,000 to 90,000 per acre, a range of 18-fold. The recorded rates of parasitism ranged from 3 to 10.6 percent; but during years of low to moderate host densities, the average rate of parasitism seemed to be as high as, if not higher than, the average rate during years when the host densities were high. Studying *E. tenebrans* in Connecticut, Arbuthnot (1955) found a higher rate of parasitism for the summer ECB brood than for the fall brood during 10 of 11 years. The data by Hill et al. (1978) also indicated a higher rate of parasitism by *E. tenebrans* during the summer than during the fall ECB brood. The density of ECB larvae during the second brood can be expected to exceed that of the first brood by severalfold (Knipling 1979, Jarvis and Guthrie 1987). Thus, available parasitism data suggest a negative rather than a positive correlation between the rate of parasitism due to *E. tenebrans* and the usual ECB density ranges.

Some insect pests tend to aggregate and then deposit their eggs in clumps or patches within small areas of their habitat rather than at random throughout the habitat. The larval hosts that develop in a few clumps or patches when the adult host population is very low may be distributed on the host plants in a manner similar to the distribution that occurs when the adult host population may be higher by 10-fold or more. Thus, if searching parasites have the capability of readily locating the infested clumps because of chemical or other cues, the probability of their finding their normal quota of hosts may be about the same whether the infested areas in the habitat are few or numerous. It will largely depend on the relative amounts of time the parasites spend searching in the infested areas and in the uninfested areas.

Thus, in principle, host-guidance mechanisms would provide the same advantage whether the hosts are distributed at random or in patches. Highly mobile parasite species that can readily locate host habitats may have little difficulty finding and parasitizing enough hosts during their lifetime to ensure viable populations even though the host population may approach extinction.

The theory that many parasites have the capability of finding their normal lifetime quota of hosts over a wide range of host densities would be consistent with the often-made observation that host density and rate of parasitism are not positively correlated. Therefore, parasite responses to kairomones are proposed as one of the main reasons for the lack of such a correlation. The theory also helps explain why so many parasite species have survived for perhaps millions of years even though host populations may often have existed at abnormally low levels for extended periods as a result of natural catastrophies. If this theory is correct in principle, even if not in absolute terms, it is a matter of profound significance for the parasite augmentation concept.

Host Discrimination

As noted, insect behaviorists recognize several steps involved in the parasitization process. The most important are host-habitat finding, host location, and host acceptance. Kairomones play a vital role in the first two steps. However, after a host is found, the parasite determines whether it is suitable for parasitizing. Prior parasitization by conspecifics (intraspecific competition) and prior parasitization by competing species (interspecific competition) are probably major factors that determine whether the host is suitable for parasitizing. Since many parasites are solitary species and only one progeny normally survives from a parasitized host, reparasitizing such hosts would be a waste of the parasites' time and resources. Not surprisingly, therefore, some parasite species have evolved capabilities to minimize intraspecific and interspecific competition. If the species cannot detect previously parasitized hosts, not only will the time and energy expended in locating such hosts be wasted but the eggs or larvae deposited and the time required to parasitize the hosts will also be wasted.

Parasites that reject previously parasitized hosts are said to exhibit discrimination. Van Lenteren (1981) reviewed the research, including his own, that had been conducted on discrimination behavior. Discrimination may be expressed in several ways. Some species leave a marking pheromone that is recognized by conspecifics and perhaps by other species. The marker could reduce the amount of host-searching time in habitats that have previously been searched. Other species recognize previously parasitized hosts on finding them and avoid reparasitizing them. However, some species apparently have not evolved discrimination behavior and will reparasitize the hosts. We have already considered the relation between probability of encountering previously parasitized hosts and the average rate of parasitism. The higher the rate of parasitism, the greater will be the probability that a host found will already have been parasitized (table 1) and, also, the greater will be the loss in average efficiency of the individual parasites.

In my view, the ability or lack of ability to discriminate is not a factor of major importance for reproductive success of a parasite species if the average rate of parasitism it causes tends to be low. Most species cause rather low average rates of parasitism. At a 20-percent rate of parasitism, the probability of encountering previously parasitized hosts will be near 10 percent. While this would represent a significant loss in efficiency, the ability or lack of ability to discriminate would seem to be a minor factor governing reproductive success, compared with all other factors. For this reason, no effort has been made in population models to assign values for the discrimination factor. This does not mean, however, that discrimination behavior is not significant in the highly competitive natural environment. Any behavior pattern that would increase the probability of reproductive success by even 1 percent would, no doubt, be highly significant. But the error, if any, that would be involved in the models proposed as representative of the coexistence patterns of various parasite-host associations included in this study would not be very large if no allowances are made for the discrimination factor. For augmented parasite populations, however, we will theoretically be dealing with very high rates of parasitism. At such rates, the probability that searching parasites will encounter previously parasitized hosts will be very high. If the augmented species exhibits discrimination behavior, the rates of parasitism could be higher than estimated. However, I take the conservative position that the encounter of previously parasitized hosts represents a complete loss in host parasitization efficiency.

Environmental Factors

The foregoing discussions, for the most part, involve biological and behavioral characteristics of the hosts and parasites and the influence they have on parasitism. However, a broad view of parasite-host associations shows that the size of any given host population is influenced much more by the total complex of environmental factors, both biotic and abiotic, than by only the behavior and efficiency of its associated parasite species. As an example, consider a host species that has a fecundity of 1,000 eggs per female. Even under the most favorable conditions, the number of adult progeny that will normally be produced is not likely to exceed an average of 10 per female parent. The accumulative mortality would therefore be approximately 99 percent. Ninety percent or more of the deaths may occur during the adult, egg, and early larval stages. If the host females each produce an average of 50 large larvae and if a given parasite species parasitizes 20 percent of the larvae, the influence of the parasite species would be minor compared with the influence of all other mortality factors. Therefore, it is apparent that the host must cope with many other hazards that are more important than a specific parasite. The parasite, too, must cope with many environmental hazards and often to a similar degree as the host. Some of the hazards for the parasite are very closely linked with the hazards that limit the reproductive success of the host.

All organisms tend to establish a steady, or characteristic, density in the environment they inhabit, as described by Huffaker et al. (1977). The populations may fluctuate widely above or below the steady density, but the fluctuations will inevitably be acted upon by natural, regulating forces with greater or lesser intensity, as required, to eventually restore the steady density. Although a parasite and its host are exposed to some of the same hazards, especially when they are living essentially as one organism, they are likely to suffer different hazards or suffer them to different extents during the time they live apart. Therefore, the relative numbers of parasites and hosts coexisting in natural populations will fluctuate to some degree from time to time. However, regardless of the normal numerical relationship between the two organisms in a stable state, I make the working hypothesis that the reproductive success of the parasite population, in terms of the average number of adult progeny produced per female parasite, will tend to equal the reproductive success of the host population on which it depends for its existence.

If a parasite population consistently had a higher reproductive success, it would, in time, completely dominate the host population. Models can be developed to show that this domination would occur within four to five generations (cycles) if the average number of adult progeny produced per female parasite exceeded the average produced per female host by as much as 25 percent each cycle. For a parasite-host association existing naturally, the ratio of adult parasites to adult hosts may vary from less than 1:10 to as high as 1:1, or higher, depending on the species. But the relative steady densities established between the parasite and host populations would indicate that their long-term reproductive successes will tend to be equal. Considerable short-term differences in reproductive success may occur from habitat to habitat, from generation to generation, or from year to year; but, as noted, when deviations do occur, natural regulating forces will act with greater or lesser intensity to restore the normal numerical relationship.

The normal or average rate of parasitism achieved by a given parasite species will depend on the relative numbers of adult parasites and adult hosts that coexist at their characteristic densities. For solitary larval parasites, it will largely depend on the ratio of adult parasites to adult hosts.

For a parasite-host association existing normally, the females in the host population will produce a certain average number of hosts of the stage preferred for parasitization by the parasites. The female parasites will find and parasitize a certain average number of hosts. As a general rule, the average number of hosts produced will have to exceed the average number parasitized by a considerable extent to compensate for the hosts parasitized. The extent can be estimated, as shown by the following examples. If parasitism by a given species averages 20 percent, the parasite-to-host ratio would be about 1:4 (20:80) and the average number of hosts produced is likely to exceed the average number parasitized by about 25 percent (100 ÷ 4). If parasitism averages 25 percent, the parasite-to-host ratio would be about 1:3 (25:75) and average number of hosts produced by the female hosts will exceed the average number parasitized by the female parasites by about 33 percent (100 ÷ 3), and so forth. This numerical relationship has evolved during thousands of years of coexistence and ensures the welfare of both organisms.

After the hosts are parasitized, the two organisms live essentially as a single organism. For oviparous species that parasitize host larvae, the eggs and all of the larval stages live within the larval host. Any environmental hazards, both biotic and abiotic, that cause the premature mortality of the host larvae will also destroy the developing parasites. The pupal stages of the two organisms generally live in close association. Consequently, they, too, share many of the same hazards. Therefore, as a general rule, I assume that for solitary larval parasites, the probability of survival to the adult stage tends to be the same for the unparasitized hosts and the immature parasites. The actual survival rates will usually diminish as the host and parasite populations grow, because of the total effect of density-dependent suppression forces. But the *relative* survival rates are assumed to remain essentially equal at all host density levels. Considerable parasite mortality that is independent of the host mortality is to be expected for some parasite-host complexes. Hyperparasitism, for example, may cause such mortality. However, for the species included in this study, I make the assumption that the survival rates tend to be approximately equal. I believe that the probable error in making this assumption is small, because the total mortalities the two organisms share during this period in their life cycles will, for most species, probably exceed by 90 percent or more the total mortalities the organisms suffer independently. However, if the normal survival rate of the developing parasites is substantially lower than that of the host, the parasite population must compensate by parasitizing a higher average number of hosts relative to the average number of hosts produced by the females in the host population. If it did not, the two populations would become hopelessly imbalanced within relatively few cycles. Elsewhere, in discussing two egg-larval parasite species, we will consider the influence of superparasitism on the rates at which the immature parasites and the unparasitized host eggs survive to the adult stage.

The theory of approximately equal survival rates for immature solitary parasites and unparasitized hosts is perhaps the most important of the theories (theories/hypotheses/assumptions) adopted for this investigation. It has several important practical implications, and rate of survival is a key parameter in all of the models developed to analyze the parasitization process for natural and augmented populations. This similarily in survival rates, together with the role of kairomones, provides a plausible reason for the lack of correlation between host densities and rates of parasitism. It also provides an important clue for estimating the normal numerical relationship likely to exist between the adult stage of a host and that of an associate parasite population if information on the normal rate of parasitism is available.

Fortunately the normal rates of parasitism caused by many species are reasonably well known. Generally, if a given species averages on the order of 10 percent parasitism, the average ratio of adult parasites to adult hosts is likely to be near 1:9 (10:90); if the average is on the order of 25 percent, the ratio is likely to average 1:3 (25:75), and so forth. Also, if a given parasite-to-host ratio is known to result in a certain rate of parasitism, the probable rate of parasitism resulting from any other ratio can be calculated. If a ratio of 1:9 causes 10 percent parasitism, two times this ratio, or a ratio of 2:9, would cause 10 + 0.10(100 – 10), or 19, percent parasitism. If a ratio of 2:9, causes 19 percent parasitism, a ratio of 4:9 can be expected to cause 19 + 0.19(100 – 19), or 34.4, percent parasitism, and so forth. Accordingly, a parasite-to-host ratio of at least 2:1 would be required for about an 80-percent rate of parasitism. In my opinion, this is the rate of parasitism to which most of our major pests would have to be subjected if we are to be reasonably sure that their populations will not continue to increase when conditions for reproduction are favorable.

Although many other factors influence the parasitization process in natural populations, the overall influence of environmental factors is no doubt the most important. Environmental factors largely determine the normal numerical relationship between a host and an associated parasite population in the ecosystem they cohabit. This numerical relationship, in turn, largely determines the proportion of the host population that will be parasitized.

Host Plants

Host plants, namely, plants fed upon by pest hosts, can influence the behavior and efficiency of parasites in several ways. Greany et al. (1977) demonstrated that fruit infested with larvae of the Caribbean fruit fly, *Anastrepha suspensa*, produce chemical products that the parasite *Biosteres longicaudatus* uses as cues to find its hosts. The parasite *Blepharipa pratensis* deposits its eggs in gypsy moth habitats where larvae have fed (Godwin and Odell 1981, Leonard 1981). These and other examples (Nordlund et al.

1981) indicate that products produced by host plants may also provide cues that guide parasites to their hosts.

Recent investigations by Lewis and Tumlinson (1988) indicate that the host-searching behavior of parasite progeny is influenced by the kind of plants on which the hosts have fed. Such a parasite response is a relatively new subject of inquiry. It indicates, however, that efficient parasitism by some species involves a learning process and that it may be necessary to condition parasites to appropriate chemical cues produced by the host insects or host plants so that the parasites will search for the hosts on the crops that require protection.

The nature of the host plants and the feeding behavior of the host insects may also influence the host-finding effectiveness of associated parasites in other ways. Host larvae feeding on foliage are likely to be more vulnerable to parasitism and predation than are host larvae that develop within stalks or fruit of the host plants. *Heliothis zea* larvae are undoubtedly more vulnerable to parasites, such as *Archytas marmoratus*, when they are developing on corn stalks or leaves than when they are developing in ears of corn. In general, stalk-boring and fruit-infesting insects are likely to be less vulnerable to biological agents than leaf-feeding species. The more vulnerable a pest is to biological agents, the higher the fecundity must be to compensate for the high mortality.

Another factor that influences host-parasitization efficiency of certain parasites is the plant-growth factor. Knipling and McGuire (1968) believed this factor to be particularly important for *Trichogramma* spp. Knipling (1971) postulated that the expanding acreage of *Heliothis* host crops during the season, as well as the plant-growth factor, is likely to have considerable influence on parasitism and predation. However, in view of the role of kairomones in guiding parasites and predators to the host or prey, these two factors may be less important than originally assumed. Also, under normal conditions, *Heliothis* populations are likely to increase as the season advances, and the increase would largely compensate for the influence of the expanding host environment. However, as will be discussed in chapter 9, if *Heliothis* larval populations are held stable or caused to decline because of high parasitism due to released *Microplitis croceipes*, the greatly expanding *Heliothis* environment may have a major negative effect on parasitism by an augmented parasite population.

Parasite and Host Fecundities

The fecundity of a parasite species (or any organism) reflects the hazards the potential progeny normally face before reaching the adult stage to continue reproduction in the next cycle. Generally, the higher the fecundity, the greater will be the accumulative mortality during the several life stages. In the final analysis, the female parents must produce enough adult progeny in their lifetime to maintain a steady density population. This is a requirement despite probable wide fluctuations in the average number of adult progeny produced per female parent during different periods in the annual or periodic cycles.

The average number of hosts that females in a population must parasitize during a given generation to ensure successful reproduction depends on the probability of survival of the immature parasites to the adult parasite stage. As a general rule, the earlier the host stage parasitized, the greater the hazards to survival before the immature parasites complete their development and, hence, the higher the average number of hosts that must be parasitized per female parent. As examples to support this general rule, we will consider two tropical fruit fly parasites that occur in Hawaii. *Biosteres arisanus*, an egg-larval parasite, parasitizes the eggs of the oriental fruit fly; and the larvae complete their development in the larger host larvae. Another related species, *B. longicaudatus*, parasitizes the large host larvae (third instar) and also completes development in the host larvae and the early pupal stage of the hosts. Therefore, the probability of survival to the adult stage would be much lower for the developing *B. arisanus* parasites than for the developing *B. longicaudatus*. As will be discussed elsewhere, I estimate that under normal conditions *B. longicaudatus* females parasitize an average of about 30 host larvae. A survival rate of 20 percent would mean six adult progeny per parent female. In contrast, I estimate that *B. arisanus* females must parasitize an average of about 300 host eggs, since only about 2 percent would be expected to survive to the adult stage. This survival rate, too, would result in six adult progeny per parent female. Estimating the relative survival rates for the two species is obviously a matter of judgment, but the estimates would seem reasonable when the accumulative mortality of the oriental fruit fly progeny is considered.

According to Vargas et al. (1984), a female oriental fruit fly is capable of depositing about 1,400 eggs. From general information obtained on adult population trends, we know that the increase rate of the population during seasons favorable for reproduction is not likely to exceed an average of 2- to 3-fold per generation. Assuming a 3-fold rate of increase, this would mean six adult progeny per female or less than 0.5 percent survival of the potential progeny. We do not know what the mortality rates are during the adult, egg, larval, and pupal stages, but all must be very high. The mortality of *B. arisanus* will be closely associated with that of the oriental fruit fly from the egg to the adult stage. The mortality of *B. longicaudatus* will be closely associated with that of the oriental fruit fly during the third larval stage to the adult stage. Later, in discussing the suppression of tephritid fruit flies by the augmentation technique, I present models based on my estimates that under favorable conditions *B. longicaudatus* females will parasitize an average of 30 large host larvae and *B. arisanus* an average of 300 host eggs. The probability of survival of the two species to the adult stage is assumed to be essentially equal. Therefore, the mortality from the egg to the large larval stage must normally be near 90 percent.

These estimates are reasonably consistent with available information on the relative fecundities of the two parasite species. According to J.E. Gilmore (former director, USDA Tropical and Fruit and Vegetable Research Laboratory, Hawaii), the average fecundity of *B. arisanus* females is near 750 eggs. In contrast, Greany et al. (1976) and Lawrence et al. (1978) reported that the fecundity of *B. longicaudatus* averages about 85 eggs per female. Therefore, the fecundity of the former species exceeds that of the latter by about 9-fold. Under comparable environmental conditions, the average number of adult progeny per female is likely to be about the same for these two parasites.

The ensuing chapters on the various parasite-host associations present the procedures and factors considered in estimating the host-parasitization capability of the parasites. In relating these estimates to fecundity, I reached the conclusion that for solitary larval parasites that parasitize the hosts directly, the host-finding or host-parasitization capability is on the order of 33 percent of the fecundity. This simple rule-of-thumb estimate is not likely to be in error by a wide margin and may be as accurate as any highly elaborate field measurements. In effect, it implies that even under favorable conditions, survival during the adult stage is not likely to exceed 33 percent. This rough estimate may be equally applicable to most hosts as well as most parasites. The accumulative mortality of all the life stages must be very high not only, as noted earlier, for the tephritid fruit flies but also for most insect pests. We might consider the *Heliothis* species as another example. The females are capable of depositing about 1,000 eggs. But even under favorable conditions, only about 1 percent survive to the adult stage. The survival rates during the different life stages may vary considerably, but if the normal survival during each of the four life stages is arbitrarily assumed to be on the order of 30 percent, each female would produce eight adult progeny. So the estimate of 33 percent survival during the adult stage would appear to be quite reliable.

Not all parasites attack their host directly. We will consider two species, *Blepharipa pratensis* and *Archytas marmoratus*, that deposit eggs or larvae on the host plants. *B. pratensis* deposits eggs, which must be ingested by host gypsy moth larvae to cause parasitism. *A. marmoratus* deposits larvae, which must then find and parasitize noctuid larvae. These behavior patterns introduce a very high additional potential hazard to the parasites. As discussed later, this hazard is reflected by the high fecundity of these two species.

By simple deduction, reasonable estimates can be made of the host-parasitization capability of parasites regardless of their behavior, their fecundity, the host stage they parasitize, or the way they cause parasitism. The host and parasite species in a host-parasite association will have comparable reproductive successes in terms of the average number of adult progeny produced per female parent. Also, if more than one parasite species depends on a given host or host complex for reproduction, on the average, all the species involved will have comparable reproductive successes in terms of the average number of adult progeny produced per female parent. The parasites and their hosts could not coexist in a balanced state unless this occurs.

General Discussion

Other factors, both known and unknown, no doubt significantly influence parasite-host relationships and the normal rates of natural parasitism. However, I feel that the most important factors have been considered. They were discussed primarily from the standpoint of their relevance to the concept of regulating pest populations by artificially increasing the number of key parasite species throughout pest-host ecosystems. There is little doubt that the role a given parasite plays as a natural control agent depends largely on the relative numbers of parasites and hosts that the forces of nature permit. Thus, we will be primarily concerned with actual and relative numbers of parasites and hosts.

In subsequent chapters, we will consider how the rates of parasitism achieved by certain parasite species are influenced when their normal numbers are increased artificially by up to 25-fold or more. There is good reason to assume that such drastic changes in the normal numerical relationships will dramatically affect the rates of parasitism and the dynamics of the host populations. The general procedure used in estimating the influence of the parasite releases is to first develop a hypothetical model that is assumed to reflect the absolute and relative numbers of parasites and hosts that coexist in typical natural populations. A key parameter in such a model is an estimate of the average number of hosts likely to be parasitized during the lifetime of the coexisting female parasites. Then, by extrapolation, estimates are made of the rates of parasitism that will result from the release of a given number of parasites and from the parasite progeny that are produced in subsequent generations.

Drastic changes in the normal numerical relationship between host and parasite populations brought about by the augmentation technique will have a major influence on all of the factors that have been discussed. Some of the estimated influences are somewhat predictable, but others are largely theoretical.

It is unrealistic to envision the parasitization process as it occurs in natural environments by considering the role of individual factors only. Their importance to the parasitization process varies with the way they interact and with the biology and behavior of the parasites. In the chapters to follow, an attempt is made to describe how the various factors interact to govern the parasitization process of the different parasite-host associations included in this study.

Introduction of New Parasites

We have thus far considered the factors that influence parasitism when the parasite and host populations coexist at their characteristic densities. Another very important aspect of natural parasitism involves the parasitism trends when new parasites are first introduced in accordance with classical biological control techniques. When the introductions are first made, the numerical and spatial relationships between the parasites and hosts are vastly different from what they will be when the new parasites eventually reach stability with the environment and the hosts.

Usually, introductions are made when the host populations are high. Often the populations have already reached their characteristic densities. Also, when the introductions are made, the parasites are usually released in inoculative numbers and only in a small proportion of the host ecosystem. In fact, hundreds of successful parasite introductions have been made under such conditions. Many introductions have resulted in satisfactory control of pests; but more often, the parasite populations have stabilized at levels that result in relatively low rates of parasitism and minimal control of the host populations. Parasitism averaging near 20 percent or less is probably representative of most newly established parasite populations. While low rates of parasitism can result in significant control and benefits that may greatly exceed the costs involved, they do not usually result in satisfactory control. Thus, we are inclined to consider such introductions as unsuccessful and categorize the parasites as rather inefficient natural control agents. From nature's standpoint, however, they could be considered highly successful. The new organisms become established and thrive in the new environment but have minimal adverse effects on the host population and other associated organisms.

If we can recognize and evaluate the major factors that permit parasite populations to establish themselves and increase in new, favorable environments but eventually stabilize at their normal densities, we will gain a better understanding of natural parasitization processes. We will also gain some idea of the number of parasites that might be required for augmentation purposes. Therefore, parasitism trends due to new introductions were also analyzed from a theoretical standpoint, keeping in mind the factors that are assumed to govern the normal relationships that prevail between populations coexisting at their respective steady densities. For the analysis, I developed various hypothetical host-population models that are similar to those described in subsequent chapters and that are assumed to represent typical, stable populations of the hosts and their associated parasites. For the models, the assumptions were made that (1) only enough of the new parasites are released to result in an initial overall ratio of 1 parasite to 1,000 adult hosts and (2) the population of the new introduction will increase and eventually stabilize at a level that will cause 20 percent parasitism. A 1:1,000 ratio for solitary larval parasites is estimated to result in about 0.1 percent parasitism. Therefore, in the process of stabilizing, the parasite population would grow about 250-fold whereas the host population would remain stable or probably actually decline to some extent. It was also assumed, as noted, that when the parasite population reaches stability, the rate of parasitism will be 20 percent. This suggests that the ratio of adult parasites to adult hosts will be about 1:4. The models are not presented, but what they reveal will be described.

Clearly, a parasite population at below-normal density has a reproductive advantage over a stable host population. No information is available on the nature and magnitude of the advantage. Therefore, they must be deduced from general observations and theoretical considerations. A key factor in the deduction is the hypothesis that the probability of survival to the adult stage is about the same for immature parasites and immature unparasitized hosts. Low-density parasite populations should have no significant influence on the relative survival rates of the immature parasites and unparasitized hosts. The survival rate of the immature parasites, as noted earlier, is closely linked with the survival rate of the host stage or stages parasitized. Therefore, the advantage to the parasite when the population is below normal must lie almost entirely with the adult stage of the parasite.

Two factors can be expected to increase the reproductive success of a parasite when its population is below normal. The rate of predation of the adults will be below normal. This will mean a longer average lifespan and an increase in the average number of hosts parasitized per female parasite in relation to the average number of hosts produced by the females in the stable host population. Also, when the rate of parasitism is below normal, intraspecific competition will be minimal. When a highly mobile parasite species is released in only a small part of the host ecosystem, the adult parasite progeny will tend to disperse throughout the host ecosystem. Therefore, until the total parasite population reaches its steady density, it will gain the same advantages as any parasite population that is below its steady density: predation and intraspecific competition will be below normal for the population as a whole. These advantages will decrease progressively, however, as the total population approaches its steady density.

To gain some idea of the magnitude of the reproductive advantage for the newly introduced parasite population, I made the assumption that so long as the parasite-to-host ratio is below normal, the female parasites will parasitize an average of 25 percent more hosts than normal during their lifetime. But the females in the stabilized host population will continue to produce near their normal number of hosts. On this basis, the parasite-to-host ratio will increase from 1:1,000 to 1:4 in about 22 generations. If the average number of hosts parasitized is 50 percent above normal, the ratio of 1:4 will be reached in about half as many generations. Therefore, if a new parasite species that has only one life cycle per year is released in inoculative numbers, about 10–20 years will probably be required for it to spread throughout the host ecosystem and reach its steady density, provided the host also has only one cycle per year. If the parasite and host have four cycles per year, the steady density would be reached in about 2.5–5 years. If the parasite has two or more cycles for each host cycle, the steady density may be reached within about 2 years. These rough estimates can be related to the performance of various introduced parasites to determine whether they are realistic. The time required for a parasite species to reach its normal maximum level will, of course, depend largely on the initial overall parasite-to-host ratio and the parasite and host distribution patterns at the time of the introduction.

The scenario described is probably an oversimplification of a more complex numerical relationship that develops before the two populations eventually stabilize. For some, if not most species, the momentum for the newly introduced population to increase may be strong enough that the population will increase beyond the normal level and cause rates of parasitism that are above normal. Above-normal rates of parasitism may, in turn, reduce the host population beyond its normal steady density. The relative numerical relationship between the host and parasite populations may be expected to oscillate before eventually stabilizing. The type of analysis described can give us a good indication of the way parasitism functions when new introductions are made to control pests and when the populations reach stability. The reproductive advantage gained over the host population by a parasite population that is below its normal density is not necessarily great per life cycle, but a rather slight advantage each cycle becomes highly significant over time. The conclusion that the advantage in reproductive success largely involves the adult stage has special significance for the augmentation technique. It tends to support the hypothesis that the ratio of the adult parasites to adult hosts will largely determine the rate of parasitism that can be achieved. As will be discussed elsewhere, it should be technically feasible to greatly increase normal parasite-to-host ratios by releasing mass-produced parasites into pest-host ecosystems.

As the ratio of parasites to hosts increases toward the stable value, the average rate of parasitism will increase progressively. In the process the steady density of the host population is likely to decline, and the decline will in turn shift the reproductive advantage toward the host population. The shift will probably be due largely to two factors. The rate of predation of the adult stage will decrease. Also, the reduced density of the immature stages will cause a reduction in the rate of predation and, hence, an increase in their survival rate. This increase will also be advantageous for the parasites, however. The newly introduced parasite population will have to compete with parasite species that are already present. The resulting adjustments and influences will likely mean that the new hazard caused by the parasite will not be fully additive to the other hazards faced by the host. In such event, the steady density of the host population will probably be reduced less than would be expected from the rate of parasitism caused by the newly introduced species.

Obviously, however, the ultimate influence a newly established species will have on the rate of parasitism and on the dynamics of the host population will depend on the relative numbers of parasites and hosts that coexist at their new steady densities. And, as previously discussed, this numerical relationship is determined by complex natural balancing forces.

In the science of pest management, several options are available that might overcome the barriers that nature imposes against the role that parasite populations can play as natural control agents. It seems unlikely that much can be done to increase the survival rates of the immature stages. Their survival rates are too closely linked with the survival rates of the host stages parasitized. The possibility of improving the efficiency of parasites by genetic selection or genetic engineering is one of the options. It may be difficult, however, to greatly increase the inherent efficiency of a species and then introduce the strain into natural populations. Ways may be found to manipulate the environment and thereby increase parasite-to-host ratios, as suggested by Huffaker et al. (1977). To have much impact, however, such environmental manipulations would have to be implemented on a large scale for highly mobile parasite and host species. Large scale manipulations would probably be costly and also could unfavorably affect other beneficial organisms.

In my opinion, by far the best way to overcome the barriers imposed by nature is to select and mass-rear well-adapted, highly-host-specific parasite species and release the adults into the pest-host ecosystems in sufficient numbers to greatly increase the parasite-to-host ratios and, hence, the rate of parasitism.

As discussed in later chapters, for many parasite-host complexes, it should eventually be technically and economically feasible to establish parasite-to-host ratios that exceed normal ratios by as much as 10- to 25-fold, provided the releases are made when the host populations are at normal low levels. If such ratios were established throughout a pest-host ecosystem, there is every reason to assume that very high rates of parasitism would result and then continue for several successive cycles, even though a number of complex biological actions and changes may come into play to minimize the influence of the abnormally high parasite populations.

In chapters to follow, suppression models are presented which depict parasite-to-host ratios that are estimated to exceed the normal ratios by up to 100-fold or more. Biologists have not had the opportunity to observe the influence such distorted ratios will have on rates of parasitism and on the dynamics of the host populations, simply because they cannot occur naturally and have never been created experimentally. But the chances seem excellent that no natural biological mechanisms exist that will prevent drastic declines in populations of hosts that are subjected to such abnormally high parasite-to-host ratios.

3 Basic Principles of Pest Population Suppression by Parasites Releases

Development of Parasite-Host Population Models

In evaluating the potential of a parasite species for augmentation purposes, the first step was to develop a basic model that can be considered representative both of the actual numbers of the parasite and its associated host coexisting in natural populations and of the normal rates of parasitism. Such a model serves several purposes: It provides a means of estimating the likely actual and relative numbers of the coexisting parasite and host during the annual or periodic cycles; and when related to the normal rates of parasitism, it will indicate the efficiency of the parasite in quantitative terms. The basic model can then be used to calculate how the rates of parasitism and the dynamics of the host and parasite populations are likely to be affected when various numbers of the parasite are released into a pest-host ecosystem.

The accuracy of the estimated number of parasites required to adequately suppress populations of the insect pests included in this theoretical appraisal hinges largely on the accuracy of the basic models and the values assigned to the different parameters. Therefore, a great deal of thought and attention were given to the assignment of values for the parameters on which the models are based.

The rationale for the estimates made for the various parasite-host associations is discussed when the potential for suppressing populations of the different pest species is evaluated. It seems desirable, however, to generally describe the procedure followed and the factors considered in developing the basic parasite-host population models. All of the models show estimated values for three key parameters pertaining to the host: (1) the number of adult hosts per unit area for each generation during the annual or periodic cycles, (2) the average number of hosts that develop to the stage preferred for parasitization and that are produced per female in each generation, and (3) the rate of survival of the unparasitized, immature hosts to the adult host stage for each generation. If realistic values are assigned to these key parameters, the models will reflect actual numbers of hosts and the rates of increase in the host populations from the normal low to the normal high densities. As noted elsewhere, all insect populations establish characteristic densities in the ecosystem they occupy. These densities are what the models are intended to portray.

Information on the actual numbers of adult insects in natural populations during the various generations is not available for most species. However, in past efforts to appraise the potential of the sterility and other autocidal techniques, I achieved considerable success in deducing realistic estimates of these numbers. Entomologists have obtained much information on the biology and behavior of all of the major pest species, including quantitative information on their fecundity and rates of survival of the overwintering stages. They have also obtained a large amount of quantitative information on egg and/or larval densities on the host plants attacked by the major pests. Such information can be used to estimate actual numbers of the various life stages that are likely to exist in normal populations. In general, enough is known about the biology and dynamics of the major pests that obviously unrealistic models can be recognized.

In contrast, estimates of the number and trends of coexisting parasite populations must be based almost entirely on deductive procedures. Virtually no information is available on actual numbers of adult parasites in natural populations. The average host-parasitization capability, in numerical terms, during the life of the female parasites and the probable survival rates of the immature stages to the adult parasite stage are unknown. Indeed, it would be very difficult and costly to estimate these values by conventional research procedures. But they are key parameters that are needed to appraise the potential of the parasite release technique when used according to the conditions stipulated in this investigation.

Even though quantitative information may be limited, reasonably good estimates can be made by deductions. The most important clue is the available information on the normal rates of parasitism caused by the more prevalent species. While available data on natural rates of parasitism are usually variable and often limited as regards the amount of data obtained throughout the seasons, typical rates of parasitism are fairly well known for many species. By using available parasitism data and relying on the theory that the rates of survival of the immature parasites and unparasitized hosts to the adult stage are of the same order, good estimates can be made of the relative numbers of adult parasites and adult hosts likely to exist in natural populations. The logic of the comparable survival theory for solitary larval and pupal parasites has been discussed elsewhere. Based on this theory, the normal rate of parasitism achieved by a given species will indicate the normal ratio of adult parasites to adult hosts in natural populations. As noted elsewhere, if parasitism averages on the order of 10 percent, the ratio should be near 1:9 (10:90). If parasitism averages 25 percent, the ratio should be near 1:3 (25:75), and so forth. If the rate of parasistism resulting from a given ratio of parasites to hosts is known with reasonable accuracy, the approximate rates of parasitism that are likely to result from various other ratios can be estimated, as previously described.

Thus, if good estimates of the number of adult hosts that normally exist in natural populations have been made, satisfactory estimates can also be made of the number of parasites that normally coexist with the host populations. Included in each parasite-host population model are assigned values for three other key parameters: (1) the actual number of adult parasites per unit area for each generation during the seasonal or periodic cycles, (2) the average number of hosts parasitized per female of the coexisting parasite population in each generation, and (3) the rate of survival of the immature parasites in the hosts to the adult parasite stage for each generation. The model will then indicate the proportion of the host population parasitized each generation and the rate of growth of the parasite population. As discussed in chapter 2, parasite and host populations generally maintain a rather consistent numerical relationship despite wide fluctuations in actual numbers of parasites and hosts during the annual or periodic cycles.

Many variations of the models were developed for the different parasite-host associations before those that seem realistic from all aspects were selected. Different values were assigned to the key parameters, and their effects on population and parasitism trends were noted. If thc assumption is made that the average number of hosts parasitized per female parasite increases significantly as the host density increases but that the *relative* survival rates of the immature parasites and unparasitized hosts remain essentially equal at all host densities, totally unrealistic relative growth rates and rates of parasitism will become evident within a few cycles. This supports the conclusion by many observers that a positive response of parasitism rate to host density is not characteristic of many parasite species.

Large errors in assumed values for the key parameters can be readily recognized as unrealistic. If the assumption is made that the average number of hosts parasitized during the life of the females is 50 percent higher than can be expected for a normal population, the rate of parasitism will approach 100 percent within four to five cycles. If the assumption is made that the average is 50 percent lower than can be expected for a normal population, the rate of parasitism will approach zero within a few cycles. (See appendix for theoretical models.) I am confident that modeling procedures can be used to quantitatively estimate the inherent host-parasitization capability of parasites within a narrow range of probable error. Indeed, the assumption of rather minor deviations in *relative* values for any of the key parameters will, within a few cycles, result in parasite-to-host ratios and rates of parasitism that are obviously not realistic. Thus, there is reason to believe that the models depicting the coexistence patterns of the various parasite-host associations included in this study are reasonably representative of typical populations. If they are, the basic models and the basic parameter values that were estimated can be used with appropriate modifications to extrapolate the changes in rates of parasitism and dynamics of the host populations that will likely occur when certain numbers of parasites are released throughout pest-host ecosystems.

Proposed Use of the Parasite Augmentation Technique

The number of parasites required for release will depend on the nature of the pest, the density of the population, and the objective. Only moderately-higher-than-normal rates of parasitism occurring with reasonable uniformity throughout the pest-host ecosystem may be adequate to prevent some pest populations from growing to levels that cause unacceptable damage to certain crops. In other cases, however, the tolerance for damage may be very low, so the pest populations will have to be greatly suppressed and maintained at very low levels to ensure satisfactory control.

This publication presents a theoretical appraisal of the potential of the parasite augmentation technique to reduce and then maintain total insect-pest populations below the level of significant damage on all crops that are attacked. For some pests, the objective may be to release appropriate parasite species for the purpose of reducing normal populations to levels that can subsequently be more readily managed by other techniques, such as those involving genetic manipulations or use of insect sex attractants. For certain pests, eradication of isolated populations may be the objective. In any event, the major goal is to appraise the potential of the augmentation technique for eventually coping with many of the more important insect pests in a highly effective, economical, and ecologically acceptable manner.

As emphasized repeatedly, in view of the dispersal range of most of the major insect pests and the mobility of the parasite species evaluated, effective regulation of total pest populations will require the release of the parasites throughout a pest ecosystem. Or if releases cannot be made against total populations, they must at least be made in areas large enough to virtually eliminate insect movement as a significant negative factor. However, normal movement of the released parasites and their progeny from habitat to habitat within the release area is considered one of the most important factors that will contribute to success of the parasite augmentation technique.

Evidence suggests that the females of virtually all host-specific parasites have the capability of parasitizing their normal lifetime quota of hosts at a wide range of host densities. Therefore, a given ratio of adult parasites to adult hosts should result in a rather constant rate of parasitism whether the host density exists well below normal or varies within the normal low to normal high range. Such a result would mean that the most efficient time to initiate an augmentation program is when the host population is at or near the lowest level. For some pests, other suppressive measures may first be necessary to reduce pest populations to levels that can be economically managed on a continuing basis by the parasite release method.

A very important conclusion reached in this investigation is that a given ratio of adult parasites to adult hosts achieved by releasing parasites in pest-host ecosystems will result in essentially the same rate of parasitism, regardless of the normal rate of parasitism the species achieves as a natural control agent. As discussed in subsequent chapters, for solitary parasite species to achieve 80 percent parasitism, I estimate that the ratio of adult parasites to adult hosts must be at least 2:1. When conditions for reproduction are favorable, nearly 80 percent parasitism each cycle is considered necessary to have reasonable assurance that populations of most of the major pest species will be suppressed or at least will not increase significantly. In new areas of spread, considerably higher rates of parasitism may be required to achieve adequate control.

The parasites should be released throughout the pest ecosystem during a period corresponding to one host cycle. For regulating pest populations, the progeny of the released parasites are more important than the released parasites. While certain natural regulating forces will act with greater intensity against abnormally high parasite populations than against normal-density populations, suppression models indicate that when initial rates of parasitism are on the order of 80 percent or higher, those forces cannot come into play in time with sufficient intensity to prevent the production of enough parasite progeny to cause severe host population suppression for several successive cycles.

Characteristics of the Parasite Augmentation Technique

A good understanding of the suppression mechanisms of different control methods in relation to the pest densities and other factors is necessary if we are to make optimum use of different methods of control either alone or in combination. In a prior publication (Knipling 1979), the suppression characteristics of various insect control methods that were then in use or under development were analyzed in considerable depth. Here, I will describe in some detail the characteristics of the augmentation technique. While most of the characteristics have already been noted in the foregoing discussions, they will be described here more specifically. Some comparisons will be made between the suppressive actions of the parasite augmentation technique and other methods of control, particularly the sterility technique. Also, the advantages and disadvantages of the methods will be presented. The characteristics described are based on the various theories, estimates, and assumptions previously discussed.

1. Among the factors that influence the efficiency of different control methods, one of the most important is the density of the pest population. The efficiency of the most commonly used control methods—including use of insecticides, pest-resistant plants, and special cultural measures—remains essentially the same whether the pest density remains constant or varies. The efficiency of other control methods—including the autocidal techniques and use of traps baited with insect sex pheromones—is inversely related to the pest density. This study indicates that the efficiency of the augmentation technique is also strongly influenced by the density of the host population. The higher the host density, the greater the number of parasites that must be released to achieve a given rate of parasitism. Therefore, the technique is unlikely to be practical after pest populations have already reached high levels. It is unlikely that the inverse relationship between efficiency of the technique and host density is strictly linear, as is assumed to be true for autocidal techniques. It seems logical that when pest-host populations reach certain critically low levels, the average host-finding capability of the parasites will begin to decline. But the

decline is assumed not to be proportional to the decline in the host population. This assumption would explain why parasites can maintain viable populations even though host populations may often exist well below normal low levels. In part, the assumption would also explain why for some parasite-host associations the release of enough parasites to cause rates of parasitism well above 50 percent will, in theory, result in progressively higher rates of parasitism with each successive generation. More will be said about this important phenomenon.

2. The augmentation technique will be highly pest specific if parasite species totally or primarily dependent on the target pest are released. Pest specificity should continue to be a major emphasis in the development and use of pest-management techniques. Pest-specific techniques not only prevent a variety of environmental hazards from developing but also protect and allow us to benefit fully from the many natural biological control agents that, together, keep populations of even the most dynamic insect pests within bounds. It must be recognized, however, that regardless of its nature, any control method that greatly suppresses a pest population will adversely affect the organisms that are totally or largely dependent on the pest for survival. The organisms may even include biological agents introduced to reduce an endemic pest population below the level of economic significance. But there is a vast difference in the ecological imbalances that will result from pest-specific methods of control and from the repeated use of broad-spectrum chemicals or cultural measures that destroy the habitats of both harmful and beneficial species.

3. When the augmentation technique is initiated, it becomes a self-perpetuating suppression measure, providing above-normal control until natural regulating forces restore the normal numerical relationship between parasites and hosts. It utilizes the host resources to produce the parasite progeny required for continuing reproduction in subsequent generations. In this very important and unique respect, the augmentation technique can be likened to an atomic breeder reaction. Against some pest hosts, the augmentation technique can theoretically be shown to have the remarkable characteristic of causing a progressive increase in rate of parasitism with each succeeding parasite generation, provided the rate of parasitism during the first generation is well above 50 percent. If the comparable survival hypothesis is correct, the higher the rate of parasitism during

cycle 1, the higher will be the parasite-to-host ratio during cycle 2. If the increase in ratio is greater than the decrease in the average number of hosts parasitized per female parasite, the rate of parasitism will increase. This increase, in turn, will result in an even higher parasite-to-host ratio during cycle 3. Theoretically, the progressively suppressive action can continue for some parasite-host associations until the parasite populations virtually displace the host populations. The highly developed capability of host-specific parasites to find hosts irrespective of the host density, because of kairomones and other host-detection signals, makes such a progressively suppressive action theoretically possible. If the action can be shown to occur for a wide range of parasite-host associations, it would represent an entirely new and powerful mechanism of suppression not possessed by any other known method of control. The action could have major implications in designing the most effective strategies for regulating insect populations.

4. If the rate of parasitism achieved during cycle 1 is below 50 percent, the rate of parasitism will not increase progressively over successive cycles for the kind of parasite-host associations included in this study. This conclusion is based on two theories: (1) the survival rates of certain solitary larval parasites to the adult parasite stage and the survival rates of the associated unparasitized hosts to the adult host stage tend to be approximately equal; and (2) the relative average numbers of hosts produced by the adult female hosts and of hosts parasitized by the female parasites in coexisting populations tend to remain constant at various host densities, at least until the maximum density is reached. Comparable survival rates of hosts and parasites to the adult stage and a fairly constant numerical relationship between hosts produced and hosts parasitized, together with the influence of kairomones for host finding, are considered to be the factors responsible for the lack of positive correlations between host densities and rates of parasitism among many parasite-host complexes. These factors interact to maintain remarkably narrow numerical relationships between host and host-dependent parasite populations despite fluctuations in actual numbers that may vary by 100-fold or more during annual or generational cycles.

5. The ability and tendency of many species of parasites to disperse for long distances virtually rule out effective and practical use of those species for augmentation releases to be made in only limited parts of the host environment. The movement of a

substantial proportion of the released population and, later, their progeny out of small release areas without the influx of comparable numbers of parasites into the release areas will greatly reduce the parasite-to-host ratios and, hence, the effectiveness of the augmentation technique. While the rate and distance of movement of parasites will vary with the species, some species are known to be capable of dispersing for many miles during a single season (Stevens et al. 1975). Therefore, the release of highly mobile parasites on a farm-to-farm or crop-to-crop basis is unlikely to be a dependable alternative to the use of insecticides or other control methods that provide immediate and effective control in the areas where they are applied. The release of certain kinds of parasites, such as *Trichogramma* spp., in specific pest-host habitats has been shown to result in substantial control, and the benefits may exceed the costs. Nevertheless, it is axiomatic that the full potential of the augmentation technique cannot be realized by making releases in areas smaller than the normal dispersal range of the parasites. (See appendix for models.)

6. While parasite movement seriously limits the effectiveness of the augmentation technique when employed in only parts of a pest-host ecosystem, such movement is considered one of the most important advantages that the augmentation technique has over other methods of control if the releases are made in appropriate numbers throughout a pest-host ecosystem and if the objective is to achieve total population suppression. It is logical to assume that a host-dependent species will tend to distribute itself somewhat proportionally to the distribution of its host. It is also logical to assume that the parasite will search diligently for the target host at all host densities. No other method of insect control has the characteristic of concentrating its suppressive action where it is most needed. When autocidal techniques are employed, both sexes are generally released. Therefore, the released males will neither selectively seek out nor be selectively sought by native females. Without such action it is unlikely that the movement of sterile insects will tend to be proportional to the distribution of the native insects. The release of only males (or females) might, however, result in seeking action that could greatly increase the effectiveness of autocidal techniques.

7. The parasite movement factor can be regarded as highly desirable or undesirable, depending on the way the augmentation technique is used. If the technique is used to regulate total pest

populations before they reach economically damaging levels, parasite movement should be highly advantageous. If the technique is used to control pests in localized habitats as the need arises, parasite movement would be highly undesirable. Unlike fast-acting chemical and biological insecticides, released parasites usually exert their suppressive action too slowly for the augmentation technique to be regarded as a satisfactory means of control after pest populations have already reached or approach unacceptably damaging levels.

Relative Efficiencies of Parasite and Sterile Insect Releases

The autocidal and parasite augmentation techniques share several characteristics. Autocidal techniques involve using adult insects that are essentially 100 percent sterile or that have been treated for optimum inherited sterility effects, namely, treated with the lowest dose of radiation that will allow them to produce the greatest number of sterile F_1 progeny. Both autocidal and parasite release techniques involve numbers—the number of the target pest in natural populations and the number of insects that must be reared and released to achieve certain levels of control. The higher the natural pest population, the greater the number of insects that must be released to achieve the required degree of control. For this reason, these techniques are not considered practical for use after the pest populations have already reached damaging levels. They must be employed as preventive measures when the pest populations exist at certain low levels. As noted earlier, suppression measures that are more efficient at high pest densities may have to be applied first to reduce populations of certain pests to a level that will make these biological techniques practical and effective.

Based on its mode of action, the parasite augmentation technique is much more effective than the sterility technique. Table 2 shows the relative efficiencies of released sterile insects and parasites in mating with or parasitizing the target pest, as estimated for different ratios of released insects to native insects. The estimates are based on the accepted characteristics of the sterility technique and the deduced characteristics of the parasite release technique. The data presented for the sterile insects are straightforward. They show that the proportion of a native pest population that will mate with sterile insects depends on the *effective* ratio of sterile insects to fertile insects in the population. If the effective ratio is 1:1, 50 percent of the native females will mate with sterile males. If the ratio is 2:1, 67 percent can be expected to mate with sterile males, and so forth. As regards parasitism by solitary species, a 1:1 ratio of adult parasites

Table 2
Relative rates of parasitism and sterility in natural pest populations as estimated for different ratios of adult solitary parasites to adult hosts and different ratios of sterile insects to fertile native insects

Ratio of sterile to fertile insects or adult parasites to adult hosts[1]	Theoretical percentage of population mating with sterile insects	Theoretical percentage of host stage parasitized
1:1	50	50
2:1	67	75
3:1	75	87.5
4:1	80	93.7
5:1	83.3	96.9
6:1	85.7	98.4
7:1	87.5	99.2
8:1	88.9	99.6
9:1	90	99.8
19:1	95	99.9+
49:1	98	99.9+
99:1	99	99.9+
999:1	99.9	99.9+

[1]These high ratios arise as a result of theoretical releases of sterile insects or parasites.

to adult hosts will result in approximately 50 percent parasitism of the host larvae. A ratio of 2:1 should result in an added 50 percent parasitism of hosts not previously parasitized, or 75 percent total parasitism. A ratio of 3:1 should result in yet another added 50 percent parasitism of hosts not already parasitized, or 87.5 percent total parasitism, and so forth. The data account for the probability that previously parasitized hosts will be encountered as readily as those not parasitized. If previously parasitized hosts are encountered, their reparasitization is assumed to be redundant. If they are not reparasitized, because of discrimination behavior, the time and energy expended in finding such hosts will still be lost.

The theoretical effects shown in table 2 are based on the assumption that the released sterile insects are fully competitive with the native insects in all respects and that the probability of sterile matings will be directly proportional to the ratio of sterile insects to fertile insects in the population. In actual practice, however, released sterile insects are not likely to be fully competitive for various reasons, including damage they sustain from radiation and from handling. The estimated influence of the parasite releases is also based on the assumption that the released parasites are of normal behavior and competitiveness and that they tend to distribute themselves proportionally to the distribution of the hosts.

According to table 2, the release of parasites will be much more effective than the release of sterile insects. The higher the ratio, the greater will be the advantage of parasite releases. At a ratio of 4:1, the parasites will theoretically be more than 3 times as effective as sterile insects during the generation of release; at a ratio of 7:1, about 18 times as effective; and at a ratio of 9:1, more than 50 times as effective.

Some factors, however, will favor sterile insects. At increasing ratios of released parasites to native host insects at a given host density, the rate of parasite predation is likely to increase and thereby reduce the average host-parasitization capability of the parasites. Some allowance is made for this reduction in the suppression models. When sterile insects are released, an increase in predation can also be anticipated. However, the increase should be equal for both the released and natural populations. Therefore, increased predation would be a positive factor for the autocidal technique and a negative factor for the augmentation technique.

As noted, the estimated relative efficiencies of the two techniques are based on the premise that the released insects will be of normal behavior and competitiveness. We know, however, that sterile insects do suffer damage, so several times the theoretical numbers required may have to be released. Radiation damage would not be involved when parasites are released. In general, however, we know relatively little about the effects of rearing and handling procedures on the behavior of parasites, and these effects will have to be investigated. The progeny produced by the released parasites should, however, be reasonably normal in behavior.

Perhaps the most important unknown factor, which will be mentioned repeatedly in connection with the estimated influence of augmented parasite populations, is the influence of abnormally low host densities on the average host-finding capability of the parasites. Under such conditions there is a possibility for this capability to decline despite the presence of kairomones and other host-detection signals. If the decline is not proportional to the decline in the number of hosts, the parasite-to-host ratios will still largely determine the effectiveness of the releases. As regards sterile insect releases, abnormally low populations of native insects would not be a negative factor if, as generally accepted, the probability of sterile matings is directly proportional to the *effective* ratio of sterile insects to fertile insects.

From a numerical standpoint, there is reason to accept the premise that parasite releases will be much more effective than sterile insect releases when the pest populations exist at normal low densities. It is logical to assume that the average host-parasitization capability of parasites will begin to decline after the host populations decline below some critical level. But it must be kept in mind that the lower the pest density, the fewer the number of parasites required to achieve very high parasite-to-host ratios. We are concerned with a game of relative numbers of pests and released insects and the relative costs of achieving the required ratios.

I estimate that the average host-parasitization capability of most host-specific parasite species will not decline more than 50 percent when the pest-host population is 90 percent below the normal low level. In accordance with table 2, a 2.5:1 ratio of adult solitary parasites to adult hosts can be expected to result in at least 80 percent parasitism of the host larvae; a ratio of 5:1, to result in at least 95 percent parasitism. Some of our pest species could be readily reduced to very low levels per acre by currently known methods of control. If the release of enough parasites to achieve a parasite-to-host ratio of 10:1, or even higher, would ensure that the pest-host population will be maintained at a low level, it may be difficult to develop a more economical or more acceptable alternative means of managing some of our major pest species.

The greatest advantage of the parasite release method over the sterile insect technique is the continuing suppression provided in subsequent cycles by the parasite progeny. In the sterility technique, releases must be made each pest generation. The releases become increasingly effective, however, if they are large enough to cause a decline in the pest population. Increasing effectiveness is a unique characteristic of the sterility technique. A given number of parasites per unit area and time is also assumed to become increasingly effective if the host population is declining, but the effectiveness might not increase after the host population reaches some critical low level. Therefore, it may be that once the population of a pest species falls below some critical threshold density, the autocidal technique will surpass the augmentation technique in effectiveness as expressed in terms of the number of insects needed for control.

The relative costs involved in rearing and releasing the insects must also be considered in appraising the relative efficiencies of the two techniques. The technology for rearing insects for autocidal control is currently more advanced for most of the pest species evaluated. The costs for rearing most solitary parasites will likely be considerably higher than those for rearing the hosts. But if the parasites are polyembryonic (more than one larva develops from one egg) or have multiparasitization behavior (deposition of more than one egg or larva per attack on the host) or can be reared in vitro, rearing costs may be lower for the parasites than for the hosts.

Thus, in comparing the relative merits and limitations of autocidal and augmentation techniques on the basis of current knowledge and theories, I have reached the conclusion that the parasite augmentation technique will be many times more effective than the sterility technique. Of the two techniques, only the sterility technique is currently in use as a means of population management. Because the sterility technique may be more advantageous under certain circumstances, the last section of this chapter is devoted to discussing how the two techniques may be used complementarily, either concurrently or sequentially, in integrated programs.

Influence of the Distribution Pattern of Abnormally Low Populations of Pest Insects on the Relative Efficiencies of Parasite and Sterile Insect Releases

A number of factors that influence the relative efficiencies of the parasite augmentation and autocidal techniques have already been considered. An important factor not yet discussed is the distribution pattern of an abnormally low population of the pest insect in the pest habitats. This factor is considered so important and has so many aspects that it will be treated as a separate topic.

Before proceeding with this discussion, it would seem important to briefly comment on the effect of pest density on the efficiencies of several categories of suppression techniques. The efficiencies of the sterile release and the parasite release techniques depend on pest density—specifically, the ratio of the adult released insects to adult native insects. The efficiency of the pheromone technique, which involves using a pheromone as pest attractant, is also pest-density dependent. In contrast, the efficiency of the insecticidal technique is not significantly influenced by pest density.

Abnormally low insect-pest populations may arise as a result of natural catastrophic events in well-established areas of infestation, in the last stages of eradication programs, or in the early stages of infestation in new areas of spread. However they arise, such low populations can be effectively eliminated or contained by any of the techniques currently in use. However, the cost effectiveness of those techniques generally decreases with decreasing pest density because of uncertainty, on the part of pest managers, about the number and distribution of the pests present. That is, since the actual pockets of infestation cannot usually be determined, pest managers may have to apply insecticides on thousands of acres of pest habitats to kill only a few hundred insects. If sterile insects are employed, millions may have to be released to inhibit reproduction of less than a thousand native insects. Traps baited with pheromone can be effective only if they are set out according to the distribution pattern of the pest population so that the pheromone bait may compete favorably against the pheromone produced by the mating insect pests.

Although the parasite release technique cannot yet be regarded as a means of eradicating or preventing the establishment of low-density pest populations, it may eventually prove to be the most effective means of doing so. The reason is that the inherent nature of host-dependent parasites is to search for and locate their hosts wherever they may be. No other method of control has this pest-seeking characteristic.

The purpose of this discussion is to consider, from a theoretical standpoint, the relative merits of sterile insect releases and releases of host-dependent parasites in coping with abnormally low pest populations distributed over a wide area. The suppression characteristics of the two techniques and the dispersal behavior of the pest species to be suppressed are key factors in the consideration.

The autocidal approach will be analyzed first. We will assume that each wild mated female is capable of producing 20 adult progeny (10-fold increase) and that these progeny represent the entire population during one cycle in a 5-acre area. The oviposition behavior of the reproducing females and the dispersal behavior of their progeny before mating occurs then become very important factors influencing the number of sterile insects that must be released per unit area to achieve adequate suppression. Since it is usually not possible to pinpoint where reproduction will occur when populations are very low, the sterile insects are distributed at random in the areas at risk. A certain minimum number will be released per unit area. For this analysis, the release rate is assumed to be 200 per acre, or 1,000 on 5 acres, during a period comparable to one pest cycle. Therefore, the overall ratio of sterile insects to native insects would be 50:1 in each 5-acre area where 20 progeny from one reproducing female exist. Trap capture data, which may provide the most practical means of monitoring the status of the program, will be used to indicate the ratio of sterile insects to native insects in the population. What the overall ratio means in terms of the degree of control will depend on the dispersal behavior of the insect under suppression.

If the females of the pest generally deposit their eggs in several acres of the pest habitats and/or the adult progeny that are produced tend to disperse into sizable areas before mating, the overall ratio of released insects to native insects caught in the traps is likely to be indicative of the effectiveness of the treatment. A sterile-to-fertile ratio on the order of 50:1 or higher would probably suggest that control is adequate. However, if the pest under suppression tends to remain in clumps or patches within the 5-acre area, an average overall ratio of even 100:1 may not be high enough to result in adequate control. Under favorable conditions for reproduction, individual females of pests like the medfly, pink bollworm, and boll weevil are likely to deposit their eggs in rather small areas, and most of the adult progeny may remain in the emergence area until mated. Most of the adult progeny from one gypsy moth egg mass are likely to emerge in a very restricted area, and the females, being flightless, will deposit their eggs in the emergence area.

The normal behavior of every pest is likely to differ; but to demonstrate the importance of the dispersal behavior of a pest species as a factor influencing the efficiency of the sterile insect technique, we will arbitrarily assume the following: The adult progeny produced by one reproducing female tend to emerge and remain in one-half acre until mated, a number of such widely distributed infestations exist in a rather large ecosystem, and the monitoring traps are set out at the average rate of one trap per 5 acres. If the released and native insects are equally responsive to the attractant used, the ratios of sterile insects to native insects in traps should average about 50:1. Obviously, the ratios will be quite variable, depending on the proximity of the traps to the infested sites. However, if the overall average ratio for traps positive for native insects is 50:1, the conclusion might be that suppression is adequate. In reality, however, for the hypothetical situation described, the *effective* sterile-to-fertile ratio in the half-acre infested sites is more likely to be on the order of 5:1. At this ratio it is unlikely that progress will be made toward eliminating the pest or preventing the infestations from increasing. Thus, higher release rates would be indicated.

If both sexes of the sterile insects are released, as they usually are, the dispersal behavior of the sterile insects is unlikely to increase the probability of sterile matings. There would be no incentive for the sterile males to selectively search for the native females or vice versa. However, if sterile males only were released and if they were capable of detecting the pheromone liberated by the native females from a considerable distance, they would likely tend to concentrate in the infested sites and increase the probability of sterile matings. To reduce rearing and release costs, scientists are investigating the possibility of genetically sexing insects in the rearing process

(Lindquist and Busch-Petersen 1987). By eliminating competing sterile females, the benefits of increased mating efficiency may equal or even greatly exceed the benefits of lower rearing and handling costs. If females actively respond to pheromones produced by males, the advantage of releasing sterile males only may not be significant.

The release of host-dependent parasites throughout pest ecosystems under the conditions described should be much more effective than the release of sterile insects. There are several sound biological reasons for this hypothesis, and they involve the behavior of both the released parasites and the host. A major weakness of autocidal techniques is monogamous mating behavior of the females under the conditions described. If, for example, a female has already mated with a fertile male and flies into an area under protection by sterile releases, the first opportunity for the sterile males to contribute to control will be against the unmated adult female progeny of the mated female. On the other hand, if parasites are routinely released in the area, they will have the opportunity to parasitize the immature progeny of the single, mated female. The criterion for estimating the relative efficiencies of parasites and sterile insects is the ratio of adult parasites to adult hosts or the ratio of released sterile insects to unmated native insects in populations. As proposed earlier in this chapter, a given ratio of adult parasites to adult hosts will result in much higher control than a comparable ratio of sterile insects to fertile insects, especially at moderate to high ratios. Therefore, if parasites can provide control by attacking the immature stages, but the sterile males can do so only after those stages have matured and then only by mating with the unmated adult females, the parasites will have an advantage equal to the intrinsic rate of increase of the species. In this case, we assume a 10-fold increase rate. Therefore, this factor alone gives the parasite release technique a 10-fold advantage if the host-finding efficiency of the parasites is not adversely affected by the abnormally low host density. Another reason why the parasite release technique should be more effective is that parasites would not be irradiated. Sterile insects, as a rule, are not fully competitive because of irradiation damage. In general, however, the most important advantage the parasite release technique will have involves the inherent behavior of parasites to actively search for host habitats and hosts. As noted, sterile insects cannot be expected to selectively search for native females if both sexes are released.

We will now analyze the influence of the distribution pattern of the abnormally low populations on the efficiency of the parasite augmentation approach. An underlying basis of this analysis is that kairomones help guide the parasites to their hosts, as detailed in chapter 2.

The assumptions are that an average of 10 female progeny originating from 1 reproducing female will be concentrated in one-half acre within a 5-acre area and that the parasites, like the sterile insects, will be released at the rate of 200 per acre during a period corresponding to 1 host cycle. Thus, a total of 1,000 adult parasites constitute the parasite population, and 20 adult hosts constitute the adult host population. Since the parasites will be released randomly in a large area, the normal movement of the parasites should tend to maintain a rather constant average population per unit area. A basic premise for this study is that the ratio of time the parasites spend searching for hosts in nonhost habitats versus the time they spend searching in host habitats is 1:10. Therefore, if 10 percent (1/2 acre) of each 5-acre area is occupied by host larvae, approximately 50 percent of the parasites' host-searching time will be in the infested habitats. While the overall host density is very low, the host density in the infested areas is not necessarily at a critical low level. Thus, if the equivalent of 500 adult parasites (50 percent of 1,000) search for hosts in the infested sites, the effective ratio of adult parasites to adult hosts would be 500:20, or 25:1. The actual numbers of parasites and hosts coexisting at any given time are not likely to exceed 167 parasites and 7 hosts (namely, one-third of each population during one cycle), but the ratio in the one-half acre is assumed to remain reasonably constant at 25:1. According to table 2, this ratio should result in more than 99 percent parasitism for solitary parasite species. But even a ratio of 7:1 should cause approximately 99 percent parasitism. Therefore, the number of parasites in the one-half-acre infested sites would exceed by more than three times the number estimated necessary to cause 99 percent parasitism. Since the host density as a whole is very low, the number of parasite progeny that would be present during the next cycle (cycle 2) will not be considered. But those present should emerge in the immediate vicinity of any hosts that might survive to reproduce in cycle 3.

The theoretical results due to the parasite releases and the sterile insect releases can be compared in numerical terms. The estimated 5:1 ratio of sterile insects to fertile insects in the one-half-acre infested sites would theoretically result in 83 percent sterile matings if the males are fully competitive (table 2). The effective parasite-to-host ratio in the infested sites would be 25:1. Since a parasite-to-host ratio of 7:1 is estimated to result in about 99 percent parasitism but the sterile-to-fertile ratio would have to be 99:1 to result in 99 percent control, the parasites would be at least 14 times more efficient. However, because the effective parasite-to-host ratio in the infested area was estimated to be 25:1, the overall efficiency of the parasites would theoretically be about 42 times that of the sterile insects. Additionally, it should be kept in mind that this comparison is based on the assumption that the reproducing adult host population numbers 20 in the infested sites. In practice, if parasites were released routinely, the ratio of parasites to adult mated hosts entering the area would be 10 times higher, so the theoretical advantage would increase by more than 400-fold.

Intuitively, the reaction to such exceptionally favorable theoretical results is likely to be that they are totally unrealistic. While such a reaction may prove to be justified, it is also quite possible that it is totally unfounded. It might be argued that if parasites are so effective, rates of parasitism approaching 100 percent should occur with noticeable frequency in natural populations. To counter such an argument, we should compare the magnitude of the numerical relationship brought about by the hypothetical conditions of the release with the magnitude of the numerical relationship that might be expected in natural populations. Consider, for example, a natural host population in a long-established environment coexisting with a host-dependent solitary parasite population that normally causes 20 percent parasitism. If, for some reason, the host population declines to an average of four adults per acre, the parasite population would probably average one per acre. In contrast to this 1:4 ratio of adult parasite to adult hosts, the overall ratio resulting from the release of parasites according to the previously described hypothetical conditions would be 1,000:20, or 50:1, but the effective ratio would be 25:1 in the one-half-acre infested area. Thus, in theory, the artificially created ratio would exceed a normal ratio under the same conditions by about 100-fold. In any natural population of a host-specific species that normally causes 20 percent parasitism, it would probably be rare to observe a parasite-to-host ratio much higher than 1:4 throughout the pest-host ecosystem.

For the hypothetical situation described, essential requirements, no doubt, are that the parasite species be well adapted to the environment and have a high degree of host specificity. The presence of other more abundant hosts that would attract the released parasites would largely defeat the objective. Also, as indicted by the investigation of Lewis and Tumlinson (1988), the parasites may have to be programmed to search for the target host.

Investigations to determine the effectiveness of using parasites to regulate or eliminate populations of some of our more important pests would be difficult and costly. But as shown in the chapters to follow, the theoretical analyses of several important pests and their associated parasites have yielded such promising results that such investigations would seem to be fully justified.

It should be emphasized that theoretical analyses of the nature described cannot be expected to yield results that are highly accurate as to details. They can, however, identify certain mechanisms and principles of pest population suppression that cannot be readily visualized without the aid of appropriate models. As the complexity of insect-pest-management technology increases, pest managers will find it increasingly important to have a good understanding of not only the biology and behavior of the pests but also the characteristics of the various suppression techniques so that they may select whichever technique is likely to be the most appropriate for any given pest condition.

Synergistic Suppression by Parasites and Sterile Insects in Integrated Releases

The advantage of releasing both parasites and sterile insects for inhibiting reproduction of pest populations has already been alluded to in previous discussions. In theory, each technique enhances the suppressiveness of the other. The result is that a given number of released insects of both kinds will be many times more effective than the release of that number of either kind alone. Integrated releases promise to be exceptionally effective when the pest populations are low. In subsequent chapters, the potential advantage of such releases to suppress specific pests will be considered. Here, we will describe how the suppressive action works and the magnitude of the advantage of an integrated release program.

That mutually potentiating (synergistic) suppression is theoretically possible when sterile insects and parasites are released concurrently for pest suppression was first proposed by Knipling (1979). Consideration was given to the evidence that oriental fruit fly larvae in Hawaii are parasitized by the egg-larval parasite *Biosteres arisanus* at a rather constant rate despite considerable variation in host-insect and host-fruit abundance during the season. To account for this evidence, the hypothesis was advanced that the rather constant rate of parasitism was due to a rather constant ratio of adult parasites to infested fruit. Thus, if the ratio of parasites to infested fruit were increased through the release of parasites, the rate of parasitism should also increase. Increasing this ratio would have an effect similar to that of increasing the ratio of sterile insects to fertile insects. Therefore, if both parasites and sterile insects were released, their suppressive effects would, in theory, be mutually potentiating. A theoretical model supported this hypothesis.

More comprehensive analyses of the numerical relationships between various parasites and hosts at various host and parasite densities have led to the conclusion, presented here, that synergistic suppression would apply to all parasite-host complexes if both host-specific parasites and sterile insects are released in the same host generation. The parasite and sterile insect techniques act independently. The sterile insects inhibit reproduction of the adults, and the parasites inhibit reproduction by parasitizing the immature host stages. The effectiveness of each technique increases as the ratio of adult parasites to adult hosts and the ratio of sterile insects to fertile insects increase. As we will see, such action, in theory, will result in a very high degree of control.

Hypothetical models will be used to show the suppressive influence of the following techniques: (1) The release of sterile insects alone, (2) the release of parasites alone, and (3) the release of both sterile insects and parasites. The insect pest and the parasite need not be identified, but several basic assumptions must be made. The pest population in the models is isolated and consists of 1 million adults during generation 1. This population is assumed to represent the normal low density for the species. If not controlled, the population will increase by 5-fold, or to 5 million adults, during generation 2.

Sterile Insect Releases Only

Reproduction by the 1 million adult insects will have to be inhibited by 80 percent if the 5-fold increase rate is to be nullified. To accomplish this inhibition by the sterility technique would necessitate the release of 4 million sterile insects that are fully competitive in all respects. Thus, the sterile-to-fertile ratio would be 4:1. In actual practice, irradiated insects are unlikely to be fully competitive. However, we assume an effective ratio of 4:1. In generation 2, then, the population would again number 1 million adults. The release of 4 million sterile insects during generation 2 should again stabilize the population.

Parasite Releases Only

The objective during generation 1 is to achieve a level of parasitism that will prevent an increase in the host population. It is assumed that 80 percent parasitism of the host larvae will have the same influence as 80 percent sterile matings. As discussed elsewhere, when the size of a host population is within its normal range, a 2.5:1 ratio of adult parasites to adult hosts is estimated to result in 80 percent parasitism of the host stage parasitized. Obtaining this ratio would require the release of 2.5 million adult parasites during generation 1. However, the parasite release technique would be much more effective than the sterility technique, because of the parasite progeny produced. Based on the comparable survival hypothesis previously discussed, a parasitization rate of 80 percent during generation 1 would result in a 4:1 ratio of adult parasites to adult hosts during generation 2. Since the adult host population would not have declined, there should be little change in the average host-finding capability of the female parasites. However, an increase in the parasite population might result in some increase in the rate of predation and a slightly lower average host-parasitization capability. For this analysis, the rate of parasitism during generation 2 is estimated to be 90 percent. In such event, the ratio of adult parasites to adult hosts during generation 3 would be 9:1. Thus, an even higher rate of parasitism could be expected during generation 3. Therefore, the parasite releases alone would be much more effective than sterile insect releases alone. However, if both techniques were employed during generation 1, the numerical relationships between parasites and hosts and between sterile and fertile insects would immediately undergo dramatic changes.

Sterile Insect and Parasite Releases During Generation 1

The biology of the pest will determine the appropriate sequence of liberating the sterile insects and parasites. If adults emerge in the spring, the program should begin with the release of sterile insects, followed by the release of parasites when the susceptible host stages appear. If the species overwinters as eggs, such as the gypsy moth, the parasite should be released first, followed by the release of the sterile insects when adult emergence begins. If reproduction is continuous, as might be expected in tropical areas, the concurrent release of sterile insects and parasites would be indicated.

In this hypothetical model, we assume the emergence of the 1 million adults during generation 1. So 4 million fully competitive sterile insects would be released first. Eighty percent of the native insects would mate with the sterile insects and, as a result, produce no progeny. The remaining 20 percent, or 200,000, of the natural population would mate and produce larval progeny. When these larval progeny begin to appear, 2.5 million parasites would be released. The ratio of adult parasites to the 200,000 reproducing adult pests would thus be 12.5:1. While the larval population will theoretically be reduced by 80 percent, it is assumed that the average host-finding capability of the parasite population will not decline proportionally to the decline in the host population, because of the highly developed host-guidance mechanism of the host-specific parasites. Based on the premise that 10:1 is the ratio of time parasites spend searching for hosts in host habitats versus the time they spend searching in nonhost habitats, an 80-percent decline in the host population below the normal low level would reduce the average host-finding capability of the female parasites by a maximum of about 30 percent. On this basis, the *effective* ratio of parasites to hosts would be 8.8:1. According to table 2, a ratio of 8.8:1 should cause about 99 percent parasitism.

When both sterile insects and parasites are released, therefore, the sterile insects would not only provide 80 percent control but also increase the actual ratio of parasites to reproducing hosts to 12.5:1 and an effective ratio of 8.8:1. This ratio would theoretically result in 99 percent parasitism. Thus, the combined effect would be 99.8 percent control. In such event, the expected population of 5 million adult hosts during generation 2 would, instead, be only 10,000. In contrast, if each technique were used alone, the adult host population would number 1 million during generation 2, a difference of 100-fold.

If 4 million sterile insects were again released during generation 2, when the host population numbers 10,000, the sterile-to-fertile ratio would be 400:1. Only 25 feral males and females would be involved in fertile matings (12.5 fertile matings). But also to be considered are the parasite progeny produced during generation 1 and their contribution to control during generation 2. The parasite progeny during generation 2 would theoretically number 990,000 if the survival rate of the immature parasites is normal. Therefore, the ratio of adult parasites to adult hosts would be 99:1, but the ratio of adult parasite to reproducing adult hosts would be about 40,000:1 during generation 2. This would be about 10,000 times the 4:1 ratio that would exist if parasites alone were released.

In summary, by integrating sterile insect and parasite releases in the manner described, the sterile insects released during generation 1 would increase the parasite-to-host ratio by 5-fold during generation 1. The parasites, in turn, would enhance the effectiveness of sterile insects during generation 2 by 100-fold. Also, the ratio of parasites to the reproducing host during generation 2 would be increased by 10,000-fold. In theory, such an integrated program would alter the actual and relative numbers of parasites and hosts so drastically that the chances of reproduction during generation 2 would be nil.

Discussion

On the basis of accepted theory and practice, we can reasonably estimate for generation 1 the contributions of the sterile insect and parasite release techniques in an integrated program of the nature described. But by generation 2, the ratio of sterile to fertile insects and the ratio of adult parasites to adult *reproducing* hosts would be so high that we enter the realm of the totally unknown in efforts to predict the influence of the parasites.

It remains to be determined what role the mechanisms of suppression described might play in future pest-management strategies. The magnitude of the synergistic effect is strongly influenced by the initial ratios achieved of released insects to native insects. Although the results of calculations will not be described in detail, if the number of sterile insects and parasites released were reduced by half, the parasite releases would increase the influence of the sterile insects only about 5-fold by generation 2. However, if the release rates are two times higher than for the model described, the synergistic influence will be further magnified manifold. It must be kept in mind, however, that the feasibility of using the two techniques will still depend on the size of the natural populations and the costs of rearing and releasing the insects. Nevertheless, knowledge of this new and potentially powerful principle of pest population suppression might eventually lead to extraordinarily effective strategies for regulating populations of many major insect pests.

4 Suppression Of Sugarcane Borer Populations

Overview

The sugarcane borer (SCB), *Diatraea saccharalis* (F.), is among the most costly insect pests in the world because of the importance of sugarcane to the economy of many countries. The borer does not rank high among the destructive pests in the United States, because sugarcane is a relatively minor crop. An appraisal of the potential of the parasite augmentation technique for controlling SCB is included in this theoretical investigation, however, for several reasons. The release of parasites for SCB control has long been considered a promising approach, and research has been undertaken on several candidate species in various parts of the world. Indeed, parasites are routinely released in sugarcane fields in certain countries for SCB control (Bennett 1969). Considerably fewer parasites are released per unit area, however, than this investigation indicates will be required to rigidly manage SCB populations. Investigators in the United States have conducted considerable research on one of the most promising of the parasites, *Lixophaga diatraeae* (LD) (fig. 2). This is a larviparous tachinid that has a high degree of host specificity, a characteristic that is advantageous for ecological reasons and for providing reasonable assurance that the released parasites and their progeny will search for the target pest species.

Field trials of LD releases conducted in the United States have generally given discouraging results (McPherson and Hensley 1976, King et al. 1981). These results are understandable, however, because most of the trials involved releases in small unisolated areas. As can be expected and as well demonstrated by Summers et al. (1976), there is considerable movement of released parasites and/or their progeny out of small release areas. When the data obtained in the field trials are viewed from a broad standpoint, however, they indicate that outstanding results might have been obtained if the investigators had maintained their rate of LD release (number of LD released per unit area) but increased the area of treatment to include the whole of the SCB ecosystem. Also, when critically analyzed, the data strongly support several of the theories of parasite-host relationships that have already been discussed.

Information available on the dynamics and behavior of SCB populations and on the biology and behavior of LD strongly suggests that this SCB–LD association can serve as an excellent model for evaluating some of the factors that govern the numerical relationship between a host population and an associated parasite population. Enough information has been obtained in the field trials of LD release to relate the results to some of the theoretical results that are presented in this chapter. Additionally, SCB can be considered representative of a wide range of stalk-boring lepidoptera that cause extensive damage to cereal and grain crops in various parts of the world. Among such lepidoptera is the European corn borer, *Ostrinia nubilalis*, one of the most important crop pests in the United States. In principle, the same factors and mechanisms that regulate the coexistence patterns of SCB and LD are likely to regulate the coexistence patterns of other stalk-boring pests and their associated parasites.

In the foregoing sections of this report, the highly complex relationships and interactions between host and associated host-dependent parasite populations were analyzed in considerable detail. Since SCB is the first host species to be considered, an effort will be made to describe, in some detail, how the theories and assumptions discussed in chapters 2 and 3 relate to the role of LD as a natural control agent and the role it might play when used for augmentation purposes.

While every parasite-host complex has certain characteristics that must be considered and evaluated, some fundamental principles govern all parasite-host relationships. One such principle is that natural regulating forces limit the number of hosts and parasites that can coexist in natural populations and, hence, the role parasites can play as natural control agents. But we will analyze the hypothesis that if the normal numerical relationship is greatly altered by artificial means, these forces cannot function with sufficient intensity and timeliness to prevent a high degree of suppression of the host populations.

Since the actual numbers of parasites and SCB that normally coexist in natural populations during the seasonal cycles are not known and would be difficult to accurately determine under field conditions, they were estimated by theoretical deductive procedures. If the estimates are reasonably accurate and are consistent with observed rates of parasitism, they will enable extrapolation of the number of parasites required to increase the rate of parasitism to a level that should result in effective control.

Figure 2. *Lixophaga diatraeae* depositing its larvae in the burrow of a sugarcane borer larva. Scientists in various parts of the world have investigated the feasibility of controlling this major worldwide pest by rearing and releasing parasites in sugarcane fields. While other species may be equally or more practical for sugarcane borer control than *L. diatraeae*, the theoretical results described in this chapter suggest that this tachinid parasite is an excellent candidate for augmentation purposes if employed in the manner proposed.
(USDA photograph.)

As argued earlier, outstanding performance from parasite releases cannot be expected unless the parasites are released in adequate numbers and at the appropriate time in isolated areas or in areas large enough to minimize the adverse effect of parasite and host movement. While considerable benefit may be realized by adding parasites of certain species to pest habitats on a farm-to-farm basis, I take the position that excessive movement of highly mobile or even moderately mobile species will make it impractical to effectively regulate pest populations in this manner.

Although this analysis involves SCB and LD, other SCB parasites may be equally or more effective for augmentation purposes. Among these may be the hymenopterous parasite *Apanteles flavipes* (Cameron), which reportedly is being used with success in Brazil (Toran and Novaretti 1978, Botelho et al. 1980). In my view, any species could be effective for augmentation purposes if it is well adapted to an SCB environment during the growing season, has a high affinity for the pest, and can be economically mass-produced. LD seems particularly promising because excellent progress has already been made on mass-rearing technology (King et al. 1979).

LD contributes relatively little to natural control of SCB in the United States (Charpentier et al. 1960, Hensley 1971). The rate of parasitism caused by self-perpetuating populations is not indicative, however, of the potential of the parasites in augmentation programs. A given parasite species may cause an average of only 10 percent parasitism; another, an average of 20 percent. All other factors being equal, the latter would doubtless be the species to use for augmentation purposes. But if the former can be mass-produced at substantially less cost, it may be the more practical to use. A conclusion reached for solitary parasite species in this investigation is that it is not their normal rates of parasitism in natural populations but, rather, the ratio of the adult parasites to adult hosts that is important in the control of pest populations by the augmentation method.

In the sections to follow, estimates will be made of the normal numerical relationship between LD and SCB populations in Louisiana and of the prevailing parasitism trend. Based on the estimated host-finding capability of a given number of parasites under a given set of conditions, estimates can then be made of the proportion of a coexisting larval host population that will be parasitized if the number of parasites is artificially increased.

Before considering the coexistence pattern of LD and SCB populations in natural environments and the theoretical influence of augmented parasite populations, the results of a parasite release experiment conducted in Florida by Summers et al. (1976) will be analyzed. The results of the experiment, although conducted on a small scale against an unisolated SCB population, strongly support the theoretical results to be presented later.

The reared parasites were held in confinement until the females were ready to larviposit. The confinement avoided the mortality that would normally occur in a natural environment during the approximately 10 days' prelarviposition period. It probably also reduced the rate of dispersion that would normally occur during the prelarviposition period in the search for food and favorable habitats. The releases were made in a 40-hectare (100-acre) sugarcane field surrounded by several thousand hectares of additional sugarcane. A total of 1,500 flies of both sexes were released per hectare during a period of about 4 weeks, which corresponds to one host cycle. Prior surveys were made to estimate larval densities and the rate of natural parasitism by LD. Natural populations of LD were absent or existed below the level of detection. Therefore, all parasitism that was recorded following the releases was attributed to the released parasites. Larval collections and larval density estimates were made at 3- to 4-week intervals in the release area and in adjacent sugarcane until the end of the season. These activities were thus conducted over a period equal to about five host and parasite cycles.

The host collections included larvae in all stages of development, so the rates of parasitism recorded do not reflect actual parasitism; that is, not all larvae had received maximum exposure to the parasites. A correction factor of about 25 percent would seem to be conservative. The rate of parasitism within the 40-hectare release area averaged 78 percent during the fourth week after releases began. If we assume a 25-percent correction factor, the average would doubtless have been about 85 percent (78 + 0.25(100 - 78)). These results can be compared with a prior theoretical estimate, by Knipling (1972), that 600 LD of both sexes per acre (1,500 per hectare) will result in 50 percent parasitism of the host larvae. Apparently the earlier estimate was too conservative by a factor of at least 2-fold. Estimates made by the procedure used in this study are considered much closer to the true efficiency.

Significantly, no parasitism was observed during the fourth week in sugarcane 100 meters from the perimeter of the release area. This observation indicates that the released flies tended to remain in the release area, probably because they were sexually mature when released. However, during the period that would correspond to generation 2 (F_1), the recorded rate of parasitism decreased in the release area but averaged 47 percent in sugarcane 100 meters from the perimeter of the release area. The actual rate, for reasons given, was probably near 60 percent. Also, parasitism averaged 4 percent in sugarcane 300 meters from the perimeter of the release area. These data clearly indicate that there was extensive movement of the F_1 progeny out of the release area. Since releases were not made in adjacent areas, comparable movement into the release area could not occur. Considering the larval density data and the size of the areas where substantial parasitism occurred during the F_1 generation, there must have been a severalfold increase in the original population and further increases in subsequent generations. This conclusion supports the basic premise that the greatest benefit from the augmentation technique will likely derive from progeny of the released parasites if conditions for reproduction are favorable.

By generation 3 (F_2), considerable parasitism was recorded up to 570 meters from the release area. This distance increased to 800 meters during generation 4 (F_3) and to 1,500 meters during generation 5 (F_4). More extensive larval host collections were made during the F_5 generation, before sugarcane harvesting began. Hosts were collected at various distances from the release area. The recorded rates of parasitism ranged from 65 to 100 percent in areas up to 1,500 meters from the release area and averaged about 80 percent. But because host larvae of all stages were collected, the true average rate of parasitism was probably higher. The rate of parasitism was 6 percent at 1,700 meters and 1 percent at 3,200 meters (approximately 1.9 miles).

The experimental results shed light on several questions that can be raised regarding the validity of the suppression model. These questions are presented later. While it is difficult to accurately relate the field data to theoretical data, some interesting and useful comparisons can be made. Taking into account the recorded host density data, rates of parasitism, and approximate acreage of the sugarcane where high rates of parasitism were recorded, I estimate that by season's end, the parasite population reached a high level in an area about 20 times the size of the 40-hectare release area. I also estimate that the parasite population grew 25- to 30-fold by season's end.

Of special significance was the observation that parasitism was 100 percent at two locations (180 larvae). Suppression models will be presented for SCB as well as other pests, and they will indicate that the rate of parasitism will be near 100 percent within several cycles after the release of enough parasites to achieve high parasite-to-host ratios. The question naturally arises for all models whether near-100-percent parasitism is possible, regardless of the number of parasites coexisting with the host population. This is a fundamental question of very great significance that needs to be answered because of its relevance to the concept of pest population regulation by the augmentation technique. On the one hand, the inclination is to assume that even when the parasite-to-host ratio is high, hosts have natural escape mechanisms that will prevent virtually 100 percent parasitism. As discussed in chapter 2, the laws of chance permit a substantial escape from parasites even when the total number of parasite-host encounters exceeds by severalfold the total number of hosts present. On the other hand, if the parasite-to-host ratio is high enough, there seem to be no biological reasons for assuming that near-100-percent parasitism is not possible. In this investigation, we will be theoretically involved with ratios of parasites to hosts that exceed normal ratios by up to 100-fold or more when host densities are abnormally low. The fact that parasitism near 100 percent is seldom observed in natural populations is undoubtedly due almost entirely to inadequate numbers of parasites rather than the ability of hosts to avoid parasitism by using some natural escape mechanisms.

The influence of parasite releases in the suppression models is based on estimates of the average host-parasitization capability of the female parasites during their lifetime under natural conditions. Direct measurement of this capability in a natural population is virtually out of the question. Indirect deductive procedures were therefore used to make the estimates. The release experiment with LD yielded data that provide some indication of the average number of host larvae parasitized by the LD females in the 40-hectare release area. There was no evidence of significant movement of the released females from the area during the first host cycle. The larval density of SCB in the 40-hectare release area on the sampling date averaged 8,000 per hectare. If we assume that the average was 20,000 per hectare during generation 1, a total population of about 800,000 larvae was present. An estimated 85 percent or approximately 680,000 of the larvae were parasitized by 30,000 female LD. This would mean that an average of approximately 23 host larvae were parasitized per female. The model depicting the estimated coexistence pattern of LD and SCB populations, which will be discussed later, is based on the assumption that during the first three cycles, the female parasites will parasitize an average of from 22 to 24 host larvae. Since these efficiency estimates for the model had been derived by deduction before calculations based on the experimental data were made, the accuracy of the deductive procedures would seem to merit considerable confidence.

Perhaps the most important question that cannot be answered is, What would have been the rate of parasitism if Summers et al. had released LD at the rate indicated but in all of the sugarcane in Florida? If all of the sugarcane had been treated, the total host population during generation 2 would have been subjected to the total F_1 progeny; the total host population during generation 3 would have been subjected to the total F_2 progeny, and so forth. Based on the equal survival hypothesis, the ratio of adult parasites to adult hosts during generation 2 would have been about 85:15, or 5.7:1. According to table 2, this should result in about 97 percent parasitism when allowance is made for intraspecific competition.

Because Summers et al. released LD in a small unisolated area, many progeny of the released parasites moved into adjacent areas. In these areas, the rate of parasitism attributable to the F_1 generation was found to be 47 percent, which I estimate would have been about 60 percent if only larvae near maturity had been collected. When the LD releases were made, the SCB host population present outside of the release area was not subject to significant parasitism. Therefore, it could increase normally. If it increased by 3-fold within a 40-hectare area bordering the release area and if half of the F_1 parasite progeny entered this area, the ratio of adult parasites to adult hosts would more likely have been about 1:1 instead of 5.7:1. I estimate that a ratio of 1:1 will result in about 50 percent parasitism. Since the rate of parasitism was probably near 60 percent, it seems likely that more than half of the F_1 parasite progeny moved out of the 40-hectare release area.

The indirect ways of relating the experimental data to the theoretical estimates are obviously subject to some error. Nevertheless, the experimental and theoretical data are surprisingly similar. This similarity is reason for confidence that the theoretical models to be presented and the values for the various parameters on which they are based are not likely to be in error by a wide margin.

The information on the distance of spread of the LD population is very important. It supports the premise that if optimum results are to be achieved by the augmentation technique, even parasite species that may have only a moderate dispersal range will have to be released throughout a pest ecosystem or at least in large areas. As theorized (chapters 2 and 3), the rate of parasitism depends largely on the ratio of parasites to hosts. Therefore, if parasites are released in only some of the host habitats, the movement of parasites and hosts and the growth of the host population in nonrelease areas are likely to result in an overall parasite-to-host ratio that will be much lower than the ratios resulting from the release of parasites in all habitats.

It should be noted that Summers et al. reached the conclusion that the parasite releases in the 40-hectare plot achieved considerable control not only in the release area but in adjacent areas. Considering the extent of the parasite movement, which was clearly indicated, and the probability that some movement of the SCB moths from the higher to the lower density areas occurred also, the results of the experiment were outstanding. However, the potential of LD releases as a means of regulating SCB populations cannot be determined until properly timed releases in adequate numbers are made throughout an SCB ecosystem. The results that can theoretically be expected from different release rates made early in the season and throughout the host ecosystem will be the subject of discussion in the sections to follow.

Natural Coexistence Pattern of Sugarcane Borer and *Lixophaga diatraeae* Populations

As described in chapter 3, the general procedure followed in estimating the coexistence pattern of a host and an associated parasite population is to first develop a model that is representative of the numbers of adult hosts and adult parasites per unit area for each generation and the rates of growth of the two populations. The important parameters for SCB are the number of adults per unit area for each generation, the average number of large larvae produced per female moth during its lifetime, and the rate of survival of the larvae to the adult stage. The values assigned to these parameters for each generation will determine the size and growth rate of the population during the season.

Information is not available on actual numbers of adult SCB existing in natural populations. But research workers and pest managers have obtained a great deal of information on larval densities for use as a guide in applying control measures and in estimating damage. Considerable information is also available on parasitism rates and trends. The important parameters for LD are the number of adults per unit area for each generation, the average number of host larvae that the adult females will parasitize during their lifetime, and the rate of survival of the immature parasites to the adult stage. The values assigned to these parameters will then determine the growth rate of the parasite population.

The model depicted in table 3 is proposed as representative of the numerical relationship between the LD and SCB populations in Louisiana. Most of the values assigned to the various parameters were estimated by deductive procedures. As noted, no reliable data exist on actual numbers of adult moths or adult LD that exist in natural populations. The average number of host larvae produced per female moth and the average number that are parasitized by the female LD are not known. Likewise, no information is available on the rates of survival of the immature parasites and unparasitized host larvae to the respective adult stages. Despite the lack of definitive information on these key parameters, I consider that the model reasonably reflects the coexistence pattern of the two organisms in natural populations.

Table 3
Basic model depicting the estimated coexistence pattern of the sugarcane borer (SCB) and the associated parasite *Lixophaga diatraeae* (LD) in Louisiana

Parameter (1 acre)	**Generation**[1]			
	1	**2**	**3**	[2]**4**
Female SCB	50	241	927	1,685
Host larvae per female (3d–5th instar)	25	24	16	6
Total host larvae	1,250	5,784	14,832	10,110
Female LD	2	9	36	95
Host larvae parasitized per female	22	24	22	16
Total host larvae parasitized	44	216	792	1,520
Percent parasitism	3.5	3.7	5.3	15
Host larvae not parasitized	1,206	5,568	14,040	8,590
Survival of parasites and hosts to adult stage, percent	40	33.3	24	over-wintering
Adult SCB, next generation	482	1,854	3,370	—
Adult LD, next generation	18	72	190	—
Increase rate of host population	4.8	3.8	1.8	—
Increase rate of parasite population	4.5	4.0	2.6	—

[1]The generations will tend to overlap as the season advances, and part of a 5th generation may occur in some areas or some years. Some host larvae and parasite larvae may go into diapause during generation 3, and others may not diapause during generation 4. However, the larval host populations (3d to 5th instars) seem reasonably representative for uncontrolled populations, and substantial damage could be expected. A survival rate of approximately 1.25 percent of the unparasitized larvae during the winter would result in similar initial adult populations for years 1 and 2. Most of the mortality of diapausing larvae is probably due to sugarcane harvest.
[2]Dashes indicate that the data are not estimated.

The estimated number of large host larvae per acre during the different generations is probably reasonably representative of typical population levels. I wish to acknowledge the assistance of S.D. Hensley (U.S. Department of Agriculture) in estimating the larval densities. The rates of parasitism shown are considered reasonably consistent with published data by Charpentier et al. (1960), Hensley (1971), and McPherson and Hensley (1976). Additionally, S.D. Hensley provided unpublished data on rates of parasitism during the early generations. These data were very helpful in the development of the model.

No doubt, the numerical relationships between the SCB and LD populations vary considerably from country to country, as well as from year to year in each country. Depending on the climate and sugarcane cultural practices, the host population levels and the trend in rates of parasitism may differ considerably from those shown. However, the most important parameters that I wished to establish were reasonable estimates of the average number of large host larvae that are normally produced by the female SCB moths and the average number that are parasitized by the female LD. In my opinion, the estimates made in relative terms are not likely to deviate from true values during the early generations by more than 25 percent. Such a deviation would be well within the range of acceptable error. The model might have shown a higher average number of host larvae produced per female moth and a lower average survival rate, or vice versa, without necessarily altering the growth rate of the host population. Likewise, the model might have shown a higher average number of host larvae parasitized per female parasite and a lower rate of survival, or vice versa, without greatly altering either the growth rate of the associated parasite population or the rate of parasitism. Whichever of these possibilities is chosen for the model, however, the *relative* numbers of host larvae produced and parasitized and the *relative* survival rates during the various cycles cannot deviate much from those assumed. Otherwise, the model will indicate obviously unrealistic host and parasite population trends and rates of parasitism. For example, if the assumption is made that the average number of host larvae parasitized per female parasite is 25 percent higher than that estimated but that all other parameter values remain as estimated, the rate of parasitism by generation 4 will exceed 40 percent. If the assumption is made that the average number of host larvae parasitized is 25 percent lower than assumed, the rate of parasitism will be less than 5 percent by season's end. Such models would be inconsistent with parasitism data.

The low number of LD estimated for the first generation is based on parasitism data for overwintered larvae obtained by S.D. Hensley. The percentage of parasitized overwintered larvae ranged from about 2 to 4 percent during 4 years' observations. These data and the fact that the rate of parasitism in the fall tends to be about 10 to 20 percent indicate that the rate of survival of the diapausing LD within the host is lower than the rate of survival of the unparasitized diapausing SCB larvae.

The normal ratio of adult LD to adult SCB moths at the beginning of the season is estimated to be about 1:25. This estimate is based on Hensley's observation that the highest rate of parasitism in overwintered SCB larvae was 4 percent. The survival rates of immature parasites and unparasitized larvae to the respective adult stages are assumed to be equal. Therefore, if the ratio of adult parasites to adult hosts is 1:25 and the estimated relative numbers of host larvae produced and of host larvae parasitized during generation 1 are typical of the normal, the rate of parasitism during generation 1 would be 3.5 percent. This rate is not inconsistent with parasitism data. However, according to data obtained by Hensley, the rate of parasitism tends to increase, especially late in the season. As discussed earlier in this report, there seems to be no positive correlation between the rates of parasitism and host densities among many parasite-host associations. I think this holds for the SCB–LD association as well. In my opinion, the increase in rate of parasitism late in the season, as indicated in the model, is due to a greater decline in the average number of host larvae produced per female SCB than in the average number of host larvae parasitized by the coexisting female LD. Such a decline in the average SCB larvae per female would seem to be a plausible explanation, because there is usually a substantial increase in the rate of egg parasitism by *Trichogramma* spp. late in the season.

According to Janes and Bynum (1941) and as discussed by Knipling and McGuire (1968) and Knipling (1979), parasitism by *Trichogramma*, which seems to be positively correlated with host-egg density, increases to a high level late in the SCB season. Parasitism rates are generally 80–90 percent. Also, predation of SCB eggs and small larvae is likely to increase when their densities increase. Therefore, the average number of large host larvae produced per female SCB would decline sharply, especially during generation 4; but the average parasitized per LD female is unlikely to decline to the same degree. Thus, a sharp increase in parasitism can be expected. The increase in rate of parasitism late in the season does not mean that the high density of larval hosts causes an increase in the average host-parasitization capability of the female LD. In fact, the model shows a decline in the average host-parasitization capability during generations 3 and 4, even though the host density has increased substantially.

Attempts were made to develop models that show an increasing rate of parasitism as the season advanced on the assumption that the individual parasites will parasitize more host larvae because of the increasing larval densities. It soon became apparent, however, that a realistic model of coexisting SCB and LD populations, using realistic values for other parameters, could not be developed on this assumption.

The hypothesis that most parasites, including LD, have the capability of finding near their normal quota of hosts at a wide range of host densities helps explain why there is no apparent correlation between host density and the rate of parasitism for many parasite-host complexes. This hypothesis has great practical significance, since it would mean that the most strategic time to release parasites is when the host density is low. Fewer parasites will be required to achieve a given ratio of parasites to hosts, which largely determines the rate of parasitism. The following relative numbers are key parameters in the coexistence model: Number of SCB larvae at the stage that is vulnerable to parasitization, number of larvae parasitized, and survival rates of the immature parasites and unparasitized host larvae to adulthood. I believe that the model presented is based on realistic values for all of the key parameters. It will be used, with some modification, to estimate the rate of parasitism and its impact on the dynamics of SCB populations when parasites are added to the host ecosystem.

Influence of *Lixophaga diatraeae* Releases

Suppression Model for High Release Rate

The parasite-host population model shown in table 3 was used to develop the suppression model shown in table 4. The adult SCB population is assumed to average 50 females per acre during generation 1, and the ratio of adult parasites to adult hosts that will result in 80 percent parasitism is estimated to be 2.25:1. To achieve this ratio 112 female LD would have to be released. Since the natural LD population during generation 1 is assumed to average 2 females per acre, the release of 112 females would increase the natural population by 56-fold. The theoretical effect of this high parasite-to-host ratio will be calculated on the basis of the following assumptions:

Table 4
Suppression model depicting the estimated rates of parasitism and the influence on the dynamics of a sugarcane borer (SCB) population if enough *Lixophaga diatraeae* (LD) are released during generation 1 to result in a 2.25:1 ratio of adult parasites to adult hosts

Parameter (1 acre)	Generation			
	1	2	3	[1]4
Female SCB	50	50	19	1.6
Host larvae per female (3d–5th instar)[2]	25	25	28	35
Total host larvae	1,250	1,250	532	56
Female LD	112	200	231	118
Ratio, adult LD to adult SCB	2.25:1	4:1	12.3:1	74:1
Larvae parasitized per female[3]	18	16	[4]10	[4]2
Ratio, parasite-host encounters to host larvae present[5]	1.6:1	2.6:1	4.3:1	4.2:1
Parasitism, percent	80	92.5	98.6	98.4
Host larvae parasitized	1,000	1,156	525	55
Host larvae not parasitized	250	94	7	1
Survival, unparasitized larvae to adults, percent[2]	40	40	45	over-wintering
SCB adults, next generation	100	38	3.2	—
Survival, immature parasites to adults, percent	40	40	45	—
LD adults, next generation	400	462	236	—

[1]Dashes indicate that the data are not estimated.
[2]The average number of host larvae produced per female SCB and the survival rates of SCB and LD to the adult stage are assumed to increase when the host population declines.
[3]A natural parasite population is estimated to have an average host-parasitization capability of 22 host larvae per female parasite. In view of the abnormally high population resulting from the releases, the average efficiency is assumed to decrease 20 percent. A further decrease is assumed when the parasite population increases.
[4]See text for discussion of the influence of host-density decline on the host-finding capability of parasites.
[5]See table 1 to calculate the rates of parasitism and escape from parasitism by using the ratio of parasite-host encounters to hosts present.

1. The parasites will be released throughout an SCB ecosystem early in the season, when the natural host population is normally low. Releases will be made at intervals during a period corresponding to one complete host cycle. To the degree practical, releases in various SCB habitats will be proportional to the host density. Thorough surveys prior to and during the release period will therefore be a vital aspect of any trial or operational program.

2. Conditions for reproduction by both the host and parasite are normal and favorable. Each female SCB moth will produce an average of 25 larvae that will reach the third- to fifth-instar stages, which are preferred for parasitization by LD. The survival rate of unparasitized larvae to the adult stage is 40 percent during the first generation. Therefore, the potential intrinsic rate

of increase of the host population during the first generation is 5-fold. If not controlled, the rate of increase of the host population in subsequent cycles will tend to diminish as the host density increases. However, for a population under suppression, the intrinsic rate of increase for the proportion of the population that reproduces will increase as the host population declines.

3. Under favorable conditions when the host and parasite populations exist within the normal size range, each female LD will find and parasitize an average of about 22 host larvae during its lifetime. However, because of the abnormally high parasite population created by the releases, the female parasite's host-finding efficiency will be reduced by 20 percent; hence, each female will parasitize an average of 18 host larvae during generation 1. The average will be reduced further as the number of adult parasite progeny increases in subsequent cycles. These are arbitrary adjustments that may or may not be adequate.

4. The female LD are capable of finding near their normal quota of hosts so long as the host density remains within a normal range. However, the average will decline somewhat when the host density declines substantially.

5. The LD females search for hosts independently. Under a given set of conditions, if 1 female is capable of finding 22 hosts during its lifetime, 4 will find 88, 100 will find 2,200, and so forth. However, the higher the proportion parasitized, the greater will be the probability that previously parasitized host larvae will be encountered. Although the females can detect previously parasitized hosts (Roth et al. 1978) and may discriminate, encountering such hosts is regarded as a complete loss in efficiency. The ratio of the total number of parasite-host encounters to the total number of hosts present determines the proportion of the host population that is parasitized one or more times and the proportion that will escape parasitism by chance (see table 1).

6. The *relative* rates of survival of the immature parasites and unparasitized host larvae to the respective adult stages are assumed to remain essentially equal at all host densities; however, the *actual* survival rates are assumed to increase with a diminishing host population.

7. Superparasitism of host larvae increases as the rate of parasitism increases. However, the survival rate of the developing

parasite is not adversely influenced by superparasitism. This assumption is in accordance with observations by King et al. (1976).

8. Each parasitized host has the potential of producing one adult parasite, although a small proportion of the superparasitized host larvae (perhaps up to 10 percent) may produce more than one adult parasite. This possible positive factor is not considered, however, in calculating the results.

Discussion of Results

The suppression model (table 4) indicates that the SCB population will be virtually eliminated by the released parasites and their progeny, produced in subsequent cycles. Since the theoretical results projected by extrapolation go far beyond any data that would support them, their validity will be open to question. Therefore, it seems important to consider the model both from a broad perspective and from a closer view in efforts to judge whether it is soundly based on biological phenomena and principles.

A typical LD population during the first SCB generation, as shown in table 3, is assumed to average two females per acre; so the parasite-to-host ratio would be 1:25. The estimated rate of 3.5 percent parasitism is consistent with observed rates of natural parasitism during the early season. The release of enough LD to result in a parasite-to-host ratio of 2.25:1 would mean the release of 56 times the number of parasites that is estimated to cause 3.5 percent parasitism. The estimate of 80 percent parasitism would therefore seem plausible when allowances are made for increased parasite predation and intraspecific competition. Then, if the potential intrinsic rate of increase is 5-fold, the adult SCB population would not change.

In accordance with the comparable survival theory, 80 percent parasitism of the host larvae would result in a 4:1 ratio of adult parasites to adult hosts during generation 2 (80:20); that is, a 77-percent increase in the ratio. The average host-finding capability of the female parasites should not change, because the host density has not changed; but the average number of host larvae successfully parasitized would be reduced because of the intraspecific competition. Also, the adult parasite population would increase substantially; and as a result, an increase in adult mortality (due to predation) would be expected. Therefore, the average number of hosts parasitized is estimated to be reduced by 10 percent. This is an arbitrary but, I think, reasonable estimate. But even with the adjustments made, the increase in ratio of adult parasites to adult host would be so large that the total number of parasite-host encounters would exceed the total number of hosts present by a ratio of 2.6:1. Theoretically, this ratio will result in 92.5 percent parasitism during generation 2. The survival rates during generation 2 are assumed to remain at 40 percent, because the host density has not changed. Thus, the ratio of adult parasites to adult hosts would be 12.3:1 (92.5:7.5) during generation 3.

The numerical relationships during generations 3 and 4 will largely govern the ultimate effect of the parasite releases. Therefore, the accuracy of the values assigned to the various parameters will determine how realistic the suppression model is. The adult SCB population, based on the 5-fold increase rate, would decline by 62 percent in generation 3. The lower adult SCB population is estimated to result in about 10 percent more larvae produced per female, or an average of 28 larvae per female. Despite this increase in the average, the total larval density would decline by about 57 percent. The adult parasite population would increase by about 13 percent, however; and the increase is estimated to cause some reduction in the average number of host larvae parasitized per female parasite. The approximately 50 percent decline in host density should cause little or no reduction in the average number of hosts parasitized, because of kairomone guidance signals. However, to be conservative, I assume the females will parasitize an average of only 10 host larvae during generation 3 compared with 16 during generation 2. Nevertheless, in view of the high ratio of adult parasites to adult hosts during generation 3, the ratio of the total number of parasite-host encounters to the total number of hosts present would be 4.3:1. Theoretically, therefore, the rate of parasitism would be 98.6 percent.

Such a high rate of parasitism during generation 3 would reduce the generation-4 adult SCB population by about 92 percent even though the survivors would have an intrinsic rate of increase of greater than 6-fold. However, enough parasite progeny would be produced to increase the ratio of adult parasite to adult hosts to about 74:1 (118:1.6).

According to the model, the larval host population during generation 4 would be about 95 percent below the normal low level. Based on the assumption of a 10:1 ratio for the lengths of time the parasites search for hosts in habitats where hosts are present and where they are absent, the parasites would parasitize 35 percent of their normal quota of hosts at such a low host density. If the normal quota is 20 larvae per female, the new quota would average 7 larvae per female. But even if it is assumed that each female will parasitize only 2 host larvae, the ratio of parasite-host encounters would still be as high as 4.2:1. This ratio would theoretically result in 98.4 percent parasitism during generation 4, or the escape of only one SCB larva per acre from parasitism. The mortality of the potential overwintering larvae due to sugarcane harvest, together with normal winter mortality, should, in such event, reduce the adult SCB population to about one pair per 100 acres of sugarcane. In theory, the population would be virtually eliminated.

Theoretical Results Versus Documented Data

It is difficult to accept the theoretical results as realistic, but there are no sound reasons for rejecting them. The accuracy of the various parameters on which the results are based may be questioned, because we obviously are operating in unchartered areas. But in light of limited documented data, we might address some of the questions that might be raised. In the experiment conducted by Summers et al. (1976), 100 percent parasitization was recorded for two locations. Thus, there would seem to be no basis for questioning the possibility of near-100-percent parasitism if the ratio of adult parasites to adult hosts is high enough. However, the density of the SCB population in the model, especially during generation 4, is probably many times lower than the natural population densities that existed at the two locations when 100 percent parasitism was recorded. On the other hand, the ratio of adult parasites to adult hosts indicated in the model is probably many times higher than the ratios that existed at the two locations.

The influence of the abnormally low host densities on the average host-parasitization capability of coexisting female parasites is totally unknown. But in theory, parasite-to-host ratios resulting from parasite releases would be unusually high. We might consider some of the estimated ratios. During generation 4 in the natural population shown in table 3, the ratio of adult parasites to adult hosts would be near 1:17. This ratio is estimated to cause 15 percent parasitism. During generation 4 in the suppression model, the ratio is estimated to be 74:1, or an increase in ratio of more than 1,000-fold. Such an abnormally high ratio might more than compensate for large reductions in host-finding efficiency because of abnormally low host densities.

It can be rationalized, as discussed in chapter 2, that most parasite species could not have survived for many thousands of years unless they are highly capable of finding hosts even when the host populations are abnormally low. As regards SCB and LD, this hypothesis can be partly supported by conditions that arose naturally in Louisiana following a severe freeze during the winter of 1961–62. It seems important to relate those conditions to the theoretical analysis that has been made in the foregoing discussion. The severe freeze reduced the population of overwintering SCB larvae to an abnormally low level. At the time, there was interest in the potential of the sterile insect technique for managing low-level SCB populations. Therefore, I asked for information on the SCB population level following the severe winter. Surveys made by members of the U.S. Department of Agriculture's laboratory at Houma, LA, indicated that the overwintered larval population was reduced to an average of about 26 per acre. If we assume a 40-percent survival of the overwintered larvae to the adult stage, the adult population would have averaged about five males and five females per acre. This natural population would be about 90 percent below the assumed normal low level. Also, if we assume that the rate of parasitism of overwintering larvae was 4 percent, which is the upper level observed by S.D. Hensley, as previously discussed, the adult LD population would have averaged about one female on 5 acres of sugarcane. At the time, there was some concern that the LD population might not survive, because of the low host and parasite densities. This concern proved to be unfounded, because the rate of parasitism returned to normal throughout the area.

For the LD to have maintained a stable population after the freeze, I estimate that each female would have had to parasitize a lifetime average of about five host larvae if the survival rate of the immature parasites to the adult stage was about 40 percent. However, if a parasite population is merely held stable for one or two cycles at a time when the host population is likely to be increasing at an abnormally high rate, the parasite-to-host ratio would become grossly distorted, and the rate of parasitism would fall to very low levels. There is no evidence that the rate of parasitism by LD fell to such levels following the freeze. This analysis suggests that the average host-parasitization capability of LD is not likely to be as low as two per female even when the SCB host density is 95 percent below the normal low density, as indicated for generation 4 in the suppression model. Therefore, the parameter values used in the suppression model may be highly conservative.

One of the most potentially significant results of this investigation is the formulation of the theory that the rate of parasitism can be expected to increase to progressively higher levels following the release of enough parasites to achieve above 50 percent parasitism during the first host cycle. The theory is based on other plausible theories discussed in chapters 2 and 3. If such a progressive increase in parasitism is found to be characteristic of certain parasite-host associations, pest managers would have a new strategy to develop and incorporate into pest population management systems; and these modified systems could have outstanding potential. If the augmentation technique increases in effectiveness as the host populations decline, even if not strictly linearly as autocidal techniques are assumed to do, the augmentation technique may be superior in several ways to the sterility technique for maintaining low pest populations or even for eradicating low-density isolated populations. In theory, it is when host populations are abnormally low, as a result of either natural causes or applied suppression pressure, that the parasite-to-host ratios can most readily be artificially raised by up to 100-fold or more. For the LD–SCB association, the ratio in natural populations during generation 3 would be on the order of 1:25. In the suppression model, the ratio is estimated to be about 12:1, or 300 times the natural ratio. During generation 4, the ratio in a natural population is estimated to be about 1:17. In the suppression model, the ratio is estimated to be 74:1, or 1,258 times the natural ratio. An analysis of the data obtained by Summers et al. (1976) indicates that the results shown in the suppression model are plausible. But even if the SCB population were suppressed only 90 percent by the end of the first season and were then reduced an additional 90 percent as a result of sugarcane harvest and winterkill, the SCB population during generation 1, year 2, would probably not exceed an average of one female per acre. Disregarding possible survivors of the natural LD population, the release of 100 females per acre would then result in a parasite-to-host ratio of 100:1. If under such conditions of low host density the female LD population parasitized an average of only one host larva, parasitism would still approach 100 percent. Certainly, the prospects seem excellent that an augmentation program along lines proposed would lead to virtual extinction of an isolated SCB population, which could then be managed by the release of relatively few parasites. However, the only way to determine the potential of LD for augmentation purposes is to undertake appropriate field experiments.

Suppression Model for Low Release Rate

The basic, coexistence model in table 3 can also be used, with suitable modifications, to calculate the influence of relatively low parasite release rates. According to Bennett (1969), LD is commonly released for SCB control in some sugarcane growing areas. However, the number of parasites released per acre or hectare is much lower than models indicate will be required for a high rate of parasitism. Also, as emphasized, it is my view that releases must be made throughout the pest-host ecosystem if the full benefit of such releases is to be realized.

The model shown in table 5 is based on a release rate that would raise the ratio of adult parasites to adult hosts to 1:2 during generation 1. For an SCB population averaging 50 females per acre, a release of 25 female LD per acre would result in the 1:2 ratio. This ratio is probably much higher than that achieved in most augmentation programs.

Consistent with previously made assumptions, the number of host larvae produced per female SCB during generation 1 is estimated to

Table 5
Model depicting the estimated rates of parasitism and the influence on the dynamics of a sugarcane borer (SCB) population if enough *Lixophaga diatraeae* (LD) are released to result in a parasite-to-host ratio of 1:2 during generation 1

Parameter (1 acre)	Generation			
	1	2	3	[1]4
Female SCB moths	50	167	454	830
Host larvae per female (3d–5th instar)	25	24	21	15
Total host larvae	1,250	4,008	9,534	12,450
Female LD	25	83	210	410
Ratio, adult parasites to adult hosts	1:2	1:2	1:2.2	1:2
Host larvae parasitized per female	20	17	15	13
Total parasite-host encounters	500	1,411	3,150	5,330
Ratio, parasite-host encounters to host larvae present[2]	.4:1	.35:1	.33:1	.43:1
Parasitism, percent	33	30	28	35
Host larvae parasitized, number	416	1,202	2,670	4,358
Host larvae not parasitized	834	2,587	6,384	8,092
Survival to adults, percent	40	35	26	over-wintering
SCB moths, next generation	334	909	1,660	—
LD, next generation	166	421	819	—
Increase rate of host population	3.3	2.7	1.8	—
Increase rate of parasite population	3.3	2.5	1.9	—

[1]Dashes indicate that the data are not estimated.
[2]See table 1 to calculate the rates of parasitism and escape from parasitism by using the ratio of parasite-host encounters to hosts present.

average 25. Each female LD is estimated to parasitize an average of 20 host larvae during generation 1. This number is approximately 10 percent lower than the average assumed for a normal population and about 10 percent higher than the average assumed for the higher release rate. In subsequent generations, arbitrary adjustments are made in the values for the average number of host larvae produced and parasitized, because of the increasing host densities. Also, the estimated survival rates are changed for the same reason. The method used to calculate the results are largely self-explanatory.

Based on the estimated host-parasitization capability, the release of 25 female parasites per acre would result in a 0.4:1 ratio of total parasite-host encounters to total hosts present. When allowance is made for the probability of the parasites encountering previously parasitized hosts, as shown in table 1, the rate of parasitism would be 33 percent. If the survival rates of the parasite larvae and unparasitized host larvae to the respective adult stages are equal, as assumed for solitary parasites, the ratio of adult parasites to adult hosts would remain at 1:2 during generation 2. Population-density-dependent suppression forces are assumed to act with slightly greater intensity against the parasite population than against the host population, because the ratio of parasites to hosts is above normal. The result would be a diminished rate of parasitism. The slight increase in parasitism during generation 4 is assumed to be due to a greater reduction in the average number of hosts produced than in the average number parasitized.

The ratio of parasites to hosts achieved by the release of 25 female parasites per acre during generation 1 would not be adequate to prevent a substantial growth of the SCB population if the intrinsic rate of increase is 5-fold. As noted, 80 percent parasitism would be needed merely to stabilize such a population. Nevertheless, if the model shown in table 5 represents the results that would be achieved, considerable benefit could be expected. The number of host larvae per acre is considerably lower than that of an uncontrolled population during the early generations. Also, since parasitized larvae are likely to cause less damage than those not parasitized, some additional control would result. Based on the efficiency estimate, however, the population would not be held to a low level; and by season's end, the number of diapausing larvae is likely to approach or exceed normal numbers. Therefore, the overwintered population during year 2 could be expected to remain near normal despite releases each year.

The suppression model involving a moderate LD release rate was developed primarily for two reasons. One reason was to demonstrate that according to the estimated host-parasitization capability of LD females and the assumed intrinsic rate of increase of SCB populations under normal conditions, such a release rate cannot be expected to rigidly manage SCB populations. Considerable benefit would theoretically be realized from such releases, however, and may well justify the costs. The other reason was to demonstrate that in theory, a parasitization rate below 50 percent will not result in an increasing ratio of adult parasites to adult hosts with each successive generation. Hence, a progressive increase in the rate of parasitization cannot be expected to occur. A progressive increase in both the parasite-to-host ratio and the rate of parasitism can be expected to occur only if the ratio of adult parasites to adult hosts during the first cycle is high enough to result in more than 50 percent parasitism. The basis for this hypothesis was discussed previously.

The factors that would prevent a progressive increase in the rate of parasitism when the initial rate of parasitism is above the normal rate but below 50 percent are assumed to apply, in principle, for most parasite-host complexes involving solitary parasites of larval and pupal hosts. Inoculative releases can be expected to result in increasing rates of parasitism but only up to the time when the parasite population reaches its normal steady density. For most species, the normal rate of parasitism stabilizes well below 50 percent. I will discuss elsewhere mechanisms that certain parasite-host complexes have evolved to prevent a progressive increase in the rates of parasitism even though normal parasitism exceeds 50 percent.

Requirements, Costs, and Benefits

Estimating the costs of new approaches to pest population management is obviously difficult until the technology proposed is fully developed and proven to be effective. Yet, theoretical results of suppression methods have no practical meaning for pest managers and growers unless they have some indication of the costs and benefits that would be involved in executing a suppression program.

The suppression model presented in table 4 indicates near elimination of an SCB population by releasing parasites early in the season. For the purpose of this cost appraisal, the SCB population will be assumed to average 200 moths per acre (100 females) and the release to consist of 500 LD adults of both sexes per acre. The SCB population will then be two times higher than that projected in the suppression model. The release of enough parasites to achieve a parasite-to-host ratio of 2.5:1 would mean an initial rate of parasitism slightly higher than that for the ratio of 2.25:1 indicated in the model.

Fortunately, substantial progress has already been made in mass-rearing LD, as the result of investigations by King et al. (1979). I estimate that eventual costs for rearing LD and releasing it at the rate of 500 per acre would total about $3 per acre. A suppression program of this nature would have to be well supported by survey and monitoring activities, especially during the first year. Administrative costs would also be involved. However, it should be possible to execute the type of program proposed at a cost of about $5 per acre during the first year. If so, the benefits, even during year 1, might well exceed the costs by severalfold.

The primary objective during the first year would be to reduce the SCB population throughout the pest ecosystem to a level that could be managed continuously thereafter at minimal cost. While the theoretical suppression model indicates near elimination of a population by the end of the first year, an assumption of only a 90-percent reduction of the SCB population by season's end and a further reduction due to normal mortality during sugarcane harvest and during the winter would mean that the surviving SCB population at the beginning of year 2 should average about one pair of moths per acre.

If the program is successful in reducing the SCB population to such a low level, two options for continuous management might be considered: (1) release SCB moths at the rate of about 200 moths per acre after treating them for optimum inherited sterility effects and (2) release LD at the rate of 200 per acre. If the SCB ecosystem is completely isolated, the autocidal technique should result in eradication. The influence of the parasite releases would have to be determined, but the possibility that they also would lead to eradication of the population should not be ruled out.

Most unisolated SCB ecosystems will likely be subject to sufficient infiltration of moths from other sources, so a continuing suppression program would be necessary. The release of parasites may be more successful than the release of irradiated moths. In my opinion, the parasites are likely to readily find and concentrate in widely scattered habitats where reproduction is occurring. While a good surveillance program may result in locating infestations before they can get out of hand, the parasites may be more effective in locating light infestations than the best detection methods now available. In continuous management programs, infestations that are found need not be eliminated. If they can merely be controlled enough to prevent the usual growth rate during the growing season, most infestations should remain so low that they would be eliminated because of natural mortality during harvest and winter.

In my opinion, an effective management program could be carried out at a cost of about $2–$3 per acre per year. If so, the benefits would likely exceed by 5- to 10-fold the losses caused by the pest under current management practices. Also, since the released parasites or genetically altered moths would be specific for the target pest, significant hazards to the environment would be avoided.

5 Suppression Of European Corn Borer Populations

The basic principles of suppressing populations of the sugarcane borer by the parasite augmentation technique should also apply to a wide range of stalk-boring lepidopterous pests, including the European corn borer (ECB), *Ostrinia nubilalis* (Hubner). ECB is among the most important of the insect pests in the United States.

Efforts to control ECB by the introduction of parasites have received major consideration, along with other efforts such as the development of host plants resistant to ECB and the development of chemicals effective against ECB (Bradley 1952). Several parasites have become established and collectively, no doubt, play a significant role in reducing damage by ECB. Like many other pests, however, the degree of control is not adequate to prevent heavy losses. This inadequacy is consistent with my view that the numerical relationships between populations of many kinds of pests and their parasites are so rigidly regulated by environmental and natural balancing factors that no single parasite, or even a complex of parasites, can reach population levels that will consistently maintain host populations below economically damaging levels. Therefore, if such insect pests are to be controlled biologically, it is my view that the populations of a key parasite species, or a complex of species, will have to be greatly increased by artificial means. Several species of ECB parasites would seem to be suitable candidates for augmentation purposes. They include *Eriborus tenebrans* (Gravenhorst), *Lydella thompsonii* (Hartig), and *Microcentrus grandii* (Goidenich) (MG). Of these, MG (fig. 3) is of special interest because of its behavior. It is a polyembryonic parasite that was first investigated by Parker (1931). The parasitized host larvae may each produce 20 or more adult parasite progeny. Most of the other species considered in this investigation are solitary larval parasites, which generally provide for the development of only one parasite progeny within a host. The typical numerical relationships between solitary parasites and their hosts will differ considerably from those between polyembryonic species and their hosts.

MG is among the most successful of the established ECB parasites (Lewis 1982, Winnie and Chiang 1982). Since a single parasitized host larva may yield many adult parasites, it may be possible to rear large numbers in a protected environment at relatively low costs compared with the costs of rearing solitary species on the insect hosts. I believe, however, that under natural conditions, the average number of adult progeny produced per female parent will be essentially equal for solitary and polyembryonic species. They merely achieve the required reproduction success in different ways.

Natural Coexistence Pattern of European Corn Borer and *Microcentrus grandii* Populations

The procedure followed in the investigation of all of the parasite-host complexes included in this publication is to first develop a model that reflects the numbers and seasonal trends of typical coexisting populations of the pest host and its associated parasite. This basic model reveals values for various parameters that can be used, with appropriate modification, for estimating the influence of parasite releases on the rates of parasitism and the dynamics of the host population.

The model shown in table 6 is proposed as representative of typical coexisting ECB and MG populations in the north-central region. Absolute values are assigned for three key parameters relating to the host: (1) the average number of adult ECB per acre in each of the two generations, (2) the average number of third- to fifth-instar larvae produced during the life of the female ECB moths (these are the stages parasitized by MG), and (3) the survival rate of the host larvae to the adult host stage. This model reflects the adult and larval host densities and the rates of increase during the season.

The parameter values for the ECB population were estimated after reviewing the ecological data published by Jarvis and Guthrie (1987) and after personal communications with W.D. Guthrie (U.S. Department of Agriculture) and H.C. Chiang (University of Minnesota). The information they supplied was very helpful in developing the model.

Like all insect populations, those of ECB vary greatly from year to year and from habitat to habitat. However, the characteristic density is estimated to be 1,000 adult ECB per acre during the first generation. The estimate that each female produces an average of 16 large larvae would mean a larval population of 8,000. The adult and larval populations are assumed to increase by 3.2-fold; so in generation 2, the adult host population would be about 3,200 and the larval population near 26,000. These populations are considerably higher than those estimated previously (Knipling 1979).

Figure 3. *Microcentrus grandii.* This polyembryonic insect parasitizes larvae of the European corn borer, one of the most costly insect pests in the United States and other countries of the world. *M. grandii* is among several parasite species that have been introduced into the United States to control the European corn borer. The various species have contributed to natural control, but not enough to prevent serious losses by this pest.

Estimates have been made of the host-parasitization capability of *M. grandii* and the numbers of the parasite that would have to be released to achieve various rates of larval parasitization. Because the parasite is polyembryonic, its mass-production on host larvae may be possible at low cost. Based on the estimated number of *M. grandii* required to achieve high rates of parasitism and on the estimated eventual costs of rearing the parasite, the prospects seem excellent that losses due to the corn borer could be prevented at one-fifth to one-tenth the cost of the losses the pest causes under present management practices. The possibility of integrating the release of the parasite and irradiated European corn borer moths as a means of regulating populations of the pest on a areawide basis is discussed.
(Photograph, courtesy of Richard L. Jones, University of Minnesota.)

Table 6
Basic model depicting the estimated coexistence pattern of natural populations of the European corn borer (ECB) and the parasite *Microcentrus grandii* (MG)

Parameter (1 acre)	Generation	
	1	[1]2
ECB females	500	1,620
Host larvae per female (3d–5th instar)	16	16
Total host larvae	8,000	25,920
MG females	250	800
Ratio, adult parasites to adult hosts	1:2	1:2
Host larvae parasitized per female	3.2	3.2
Total host larvae parasitized	800	2,560
Percent parasitism	10	10
Unparasitized host larvae	7,200	[2]23,360
Average parasite larvae per parasitized host	20	20
Total parasite larvae	16,000	51,200
Survival of unparasitized hosts to adults, percent	45	over-wintering
Total ECB adults, next generation	3,240	—
Survival of MG larvae to adults, percent	10	—
Adult MG, next generation	1,600	—
Increase rate of ECB population	3.2	—
Increase rate of MG population	3.2	—

[1]Dashes indicate that the data are not estimated.
[2]If 60 percent of the unparasitized larvae survive to become mature diapausing larvae and 7 percent of the diapausing larvae survive to the adult stage, the 1st-generation population of female ECB's in year 2 would be similar to that in year 1.

An important parameter is the estimated survival rate of the ECB larvae to the adult stage. The large larvae and pupae are well protected in their burrows, so the survival rate is estimated to be about 40–50 percent. Obviously the various parameter values, will deviate considerably from generation to generation and from year to year. However, the model is considered reasonably representative of average conditions.

The key parameters for the coexisting MG population include (1) the number of adult MG per acre during the two generations, (2) the average number of ECB larvae parasitized by the females during their lifetime and (3) the estimated survival rate of the immature parasites to the adult stage. Information on these important parameters is not available, and accurate data would be very difficult to obtain by field investigations. Therefore, the quantification of these parameters was based largely on theories and deductions. An estimate of the average host-parasitization capability of the female parasites under normal conditions is the most important parameter. The procedure used and the factors considered in estimating the key parameters will be described briefly.

The theory of comparable survival rates for the immature stages of a solitary parasite species and the unparasitized larvae of the associated host to the respective adult stages was one of the most useful theories for estimating the ratio of adult parasites and adult hosts likely to coexist in natural environments. The ratio will depend on the normal rates of parasitism.

The theory will not apply, however, for a polyembryonic species, such as MG. Parker (1931) found that the parasitized ECB larva has the potential to yield an average of about 20 MG progeny. Shyu's (1981) laboratory study indicated that the average may be considerably higher. Such data—together with the basic theory that the reproductive success of a parasite species totally dependent on a given host for reproduction will, on average, equal the reproductive success of the host—allow realistic values to be estimated for other key parameters. Thus, the MG and ECB populations are assumed, on average, to increase at essentially the same rate. The rationale for this theory was discussed in chapter 2.

Therefore, if a typical ECB population during generation 1 increases by 3.2-fold, the MG population during generation 1 would also increase by 3.2-fold. Apparently no data are available on the average number of adult parasite progeny that develop per host larva in natural populations (personal communication with R.L. Jones, University of Minnesota). Consequently, the coexistence pattern was estimated on the assumption that naturally parasitized larvae have the potential of producing an average of 20 adult progeny. The normal survival rate of the immature parasite progeny then becomes an important parameter. Considering the behavior of the parasitized host larvae and the developing immature parasites, it is my opinion that the survival rate is normally very low. Various assumed survival rates were considered, ranging from 5 to 20 percent. A survival rate of 10 percent is assumed to be closest to the probable normal survival rate. On this basis, the MG female would have to produce an average of 64 larval progeny. Then, if 10 percent normally survive to the adult stage, the females would each produce an average of 6.4 adult progeny, which represent a 3.2-fold rate of increase. Since each parasitized host has the potential to yield 20 adult progeny, the MG female would have to parasitize an average of 3.2 host larvae.

There is no way of knowing how accurate the assumed 10-percent survival factor is. But if each female parasitizes an average of only 2 host larvae, the rate of survival of 40 potential progeny would have to be 16 percent. If it parasitizes an average of 5 host larvae, the survival rate of 100 potential progeny would have to be as low as

6.4 percent. The accuracy of the estimated rates of parasitism that will result from the release of a given number of MG into ECB ecosystems hinges on the accuracy of both the estimated average number of progeny produced per parasitized larvae and the assumed survival rate. In my opinion, the true host-parasitization efficiency of the parasites will fall within the range of two to five hosts per female parasite. If the average is as low as two per female, MG releases would not necessarily be ruled out as a practical means of suppressing ECB populations. But the number of adults released would have to be about two times the number estimated for the simulation suppression model that will be presented later. I believe the probability is very high that under normal conditions, the female MG will be capable of parasitizing an average of three to four host larvae.

If we accept the estimate that the MG female normally parasitizes an average of 3.2 host larvae, calculations can be made for the number of MG adults that coexist with a typical ECB population, as shown in table 6. This number will depend on the normal rate of parasitism. If parasitism averages 10 percent, 800 of the 8,000 ECB larvae present during generation 1 would be parasitized. If the MG female normally parasitizes an average of 3.2 host larvae, the female MG population would be 250 per acre (800 ÷ 3.2 = 250). If normal parasitism is higher or lower, the ratio would change, but the average host-parasitization capability of the parasite would change little, if any.

Two procedures can be used to estimate the proportion of a host population that will be parasitized. If the average number of hosts that females of a species are capable of parasitizing during their lifetime is known within a reasonable range, calculations can be made of the number of hosts that will be parasitized by any given number of parasites when added to the host ecosystem. However, allowance must be made for the probability that previously parasitized hosts will be encountered. The other procedure is based on the ratio of adult parasites to adult hosts. If the rate of parasitization resulting from a given ratio is known, the rate of parasitism resulting from an increase in that ratio can be estimated. For example, if a ratio of 1:2 results in 10 percent parasitism, a ratio of 2:2 (or 1:1) should result in 10 + 0.10(100 - 10), or 19 percent, and so forth. The rates of parasitism calculated by both procedures are generally similar.

In the following section, the influence of MG releases on the rate of parasitization and on the dynamics of an ECB population will be calculated largely on the basis of the information in table 6.

Influence of *Microcentrus grandii* Releases

Suppression Model

The suppression model depicting the influence of MG releases in ECB ecosystems is based on the following assumptions:

1. The ECB population exists essentially as depicted in table 6.

2. During the period corresponding to generation 1, 6,000 MG females will be released per acre. The releases will be made in areas substantially larger than the normal flight range of the parasite and ECB moths.

3. To the degree practical, MG releases will be proportional to the ECB density. Therefore, thorough surveys will be required in the program.

4. The host-parasitization capability of the MG females in natural populations averages 3.2 host larvae. However, due to the high population resulting from the releases, the average is 2.6 per female. This average represents a 20-percent adjustment based on the assumption that natural predation will increase as the parasite population increases and that therefore the reproductive life and average host-parasitization rate of the female parasites will be reduced.

5. The parasites search for hosts independently. Thus, if 1 female has the capability to parasitize 2.6 host larvae during its life time, 1,000 will parasitize 2,600; 5,000 will parasitize 13,000; and so forth. The proportions of the host larvae that will be parasitized, superparasitized, and spared parasitization will depend on the probability of encounter between adult MG females and previously parasitized hosts. The data in table 1 are used to calculate rates of parasitism from ratios of the number of parasite-host encounters to the number of hosts present.

6. Each parasitized host larva has the potential to produce 20 adult MG progeny. However, the rate of survival of the immature parasites to the adult parasite stage, as noted, is assumed to be 10 percent.

The simulation suppression model is shown in table 7. The method of calculating the influence of the releases is largely self-explanatory.

Table 7
Suppression model depicting the estimated influence of *Microcentrus grandii* (MG) releases in European corn borer (ECB) ecosystems like those depicted in table 6

Parameter (1 acre)	Generation	
	1	[1]2
ECB females	500	270
Large host larvae per female	16	[2]18
Total host larvae	8,000	4,860
MG females released (generation 1 only)	6,000	6,800
Ratio, adult MG to adult ECB	12:1	25:1
Average host larvae parasitized per female	2.6	2.6
Total parasite-host encounters	15,600	17,680
Ratio, parasite-host encounters to host larvae present[3]	1.9:1	3.6:1
Parasitism, percent	85	97
Host larvae parasitized	6,800	4,714
Host larvae not parasitized	1,200	[4]146
Survival, unparasitized larvae to adults, percent	44.5	overwintering
Adult ECB, next generation	540	overwintering[4]
Average MG larvae per parasitized host	20	—
Total MG larvae	136,000	—
Survival, MG larvae to adults, percent	10	—
Adult MG, next generation	13,600	overwintering[5]

[1]Dashes indicate that the data are not estimated.
[2]In view of the greatly reduced ECB population below normal, the average number of host larvae produced per female ECB is assumed to increase.
[3] See table 1, to calculate the rates of parasitism and escape from parasitism by using the ratio of parasite-host encounters to hosts present.
[4]If 60 percent of the unparasitized larvae survive to become mature diapausing larvae and 10 percent of these survive to the adult stage, the adult ECB population would average about 9 per acre during year 2.
[5]In view of the high rate of parasitism during generation 2, the ratio of adult MG to adult ECB during year 2 should be about 100:1.

Discussion of Results

The release of the 6,000 MG females (and 6,000 males) proposed would theoretically result in a 12:1 ratio of adult parasites to adult hosts. This ratio is 24 times the 1:2 ratio estimated to result in 10 percent parasitism. When allowance is made for the abnormally high parasite density and the encounter of previously parasitized hosts, the rate of parasitism resulting from the released parasites would be about 85 percent. At this rate and the potential of the host population to increase at a rate of 3.2-fold, a 46-percent decrease in the ECB population would be expected in generation 2. This decrease is assumed not to cause a significant change in the average host-parasitization capability of the MG females. But the decrease is assumed to increase the average number of host larvae produced by the ECB females by about 10 percent, or to an average of about 18 per female.

Based on the relative rates of survival of the parasite larvae and unparasitized host larvae to the respective adult stages, the ratio of MG adults to ECB adults would increase to 25:1 during generation 2. The rate of parasitism during generation 2 is estimated to increase to 97 percent. The possibility that parasitism could reach such a high level can certainly be questioned. However, I believe that reasonably realistic values have been assumed or estimated for the parameters on which the model is based. While parasitism approaching 100 percent is seldom observed in natural populations of most insects, the ratio of parasites to hosts in natural populations cannot be expected to even approach the ratios that could be achieved by artificial means. We can be reasonably certain that ECB populations have never been subjected to numbers of MG that would increase the normal parasite-to-host ratio by 25-fold or more throughout the host ecosystem. Therefore, until appropriate experiments are conducted, we have no way of knowing what the results might be. But such a distorted ratio should greatly increase the rate of parasitism.

As discussed in chapters 2 and 3, the rate of parasitization caused by a natural population of a parasite species is not a reflection of its host-finding capability in numerical terms. Rather, it is a reflection of the relative numbers of parasites and hosts that natural environmental forces permit in natural populations. The significance of this theory should not be considered casually or lightly. There is relatively little we can do to increase the inherent host-finding capability of a parasite species. It would probably be a major research achievement to develop a parasite strain, through selective breeding, that is two times as efficient in host finding than a normal strain. However, we have or could have the technology to increase by 25- to 50-fold the normal numerical relationship that exists between a host and an associated parasite population. It remains to be determined what influence such a drastic increase in the normal numerical relationship between MG and ECB populations would have on rates of parasitism and on the dynamics of ECB populations.

As noted earlier, the solitary parasites *E. tenebrans* and *L. thompsonii* may also be effective when released in ECB ecosystems. Using these species to achieve results similar to those achieved with MG, I estimate that the initial ratio of adult parasites to adult

ECB would have to be about 2.5:1. The number that would have to be released is only about one-fifth the estimated number of *M. grandii* required. However, if the parasites have to be reared on the host larvae, the cost of rearing the parasites is likely to exceed the cost of rearing *M. grandii* by 10-fold or more. If in vitro rearing procedures could be developed for the solitary species, they may prove to be more practical for augmentation purposes than *M. grandii.*

Requirements, Costs, and Benefits

The theoretical results projected in the suppression model are very impressive. The host population during year 1 would be reduced by about 99 percent. If the relative survival rates of parasite and host larvae during the winter remain about the same, the parasite-to-host ratio of adults would be about 99:1 during generation 1 in year 2. However, no effort is made to estimate results during year 2. While the theoretical results may be impressive, they have no practical significance unless the number of insects required is related to the probable costs of rearing and releasing them, and these costs are, in turn, related to the probable benefits. The mass-rearing costs that would be involved are unknown but can be estimated with considerable confidence because of the experience gained in recent years by scientists on the mass-rearing of a wide range of insects. These estimates could then be the basis for estimating the benefits.

I estimate that ECB larvae could eventually be mass-produced at a cost of $5 per 1,000. If, as previously noted, each host is assumed to be capable of producing 20 MG progeny, the mass-production costs for MG would be about 25 cents per 1,000. The simulation model is based on the release of 6,000 MG females per acre, or 12,000 of both sexes. Thus, the cost of the parasites would be about $3 per acre.

More conservatism would seem desirable, however, in projecting theoretical results of simulation suppression models to results that might be expected in practical programs. Therefore, in estimating the costs and requirements for suppressing ECB populations by MG releases, a more realistic assumption may be that two times as many parasites would have to be released for adequate suppression. On this basis, the per acre cost for the parasite would be about $6. All other costs would probably increase the costs during year 1 to about $10 per acre. In such event, the benefits may equal the costs even during year 1.

The ultimate objective, however, would be an ecologically acceptable means of maintaining ECB populations below economically damaging levels at minimum annual costs. While the suppression model indicates that the ECB population would be reduced by 99 percent to about 10 adults per acre during year 2, a more conservative estimate might be on the order of 95 percent suppression, even with two times the number of parasites shown in the suppression model. Therefore, the ECB population during generation 1 of year 2 is assumed to exist at an average level of about 50 adults per acre. The influence of overwintering parasites will not be considered.

If an ECB population is reduced to an average of 50 adults per acre, three options for further suppression during year 2 are proposed: (1) release enough MG adults (several thousand, perhaps) to ensure rates of parasitism that would at least prevent an increase in the ECB population during year 2; (2) release enough moths treated for optimum inherited sterility effects to achieve a 10:1 ratio of treated insects to native insects so that the ECB population would be held stable or further suppressed; and (3) release about half the number of both parasites and partially sterile moths. Based on the analysis discussed in chapter 3, option 3 should be the most effective means of suppression. I believe that the year-2 costs would not exceed $5 per acre. In such event, the benefits during year 2 may exceed the costs by severalfold.

The full benefits of the envisioned program may not be realized, however, until after year 2. If the year-2 population were merely held stable at 50 adults per acre, natural control during the winter should reduce ECB populations to an average level of 5–10 adults per acre by year 3. Too little is now known about the efficiency of MG or any other parasite species against very low host populations to allow an estimate to be made of the influence of continuing parasite releases. However, if a parasite-to-host ratio on the order of 100:1 would maintain ECB populations at economically nondamaging levels, as few as 1,000 adult parasites per acre each year may be adequate for continuing control. Another option—the release of from 50 to 100 genetically altered moths per acre—should maintain the ECB populations below levels that would cause significant damage during year 3 and the years following.

It is generally accepted that the sterility technique becomes increasingly effective as the pest population declines. Therefore, the routine release of irradiated moths may be the most practical procedure to consider at this time. If the natural population is virtually eliminated from most of a large area, the routine release of

50–100 moths per acre in 50- to 100-mile-wide barrier zones protecting areas subject to infiltration by ECB moths from untreated areas may be adequate to prevent the reestablishment of damaging populations. A thorough surveillance program in all areas would be necessary, however, to detect and apply appropriate suppressive measures against insipient threatening populations. After complete dominance of the ECB population is achieved, an annual expenditure ranging between $1 to $2 per acre may be adequate to continue regulating ECB populations on a regional scale. If so, the benefits may exceed costs by 5- to 10-fold.

The pest management community and the agricultural industry may not now be prepared to consider undertaking the type of program I have described. The initial costs would be high, and many technical and operational problems would have to be resolved. The ideal solution to the ECB problem may be to develop corn varieties that are virtually 100 percent resistant to the pest. But there is no assurance that resistant varieties alone will completely solve the problem. The theoretical appraisal I have presented indicates that when supplemented by an appropriate autocidal approach, the parasite augmentation approach may allow us to solve this costly pest problem effectively and at relatively low cost without harming the environment. Like the dynamics of many of our other major insect pests, the dynamics of ECB is such that an attack on only portions of the total population, no matter how intensive or prolonged, will not resolve the problem. On the other hand, this analysis indicates that by applying moderate but consistent suppression pressure against total populations in an organized and coordinated manner and also by taking full advantage of natural control factors, we may be able to maintain ECB populations well below economically damaging levels at costs that would be only a fraction of the cost of damage the pest causes under current management practices.

6 Suppression Of Tephritid Fruit Fly Populations

Overview

Tephritid fruit flies rank high among the world's most important insect pests. A number of species are involved, and they attack a wide range of high-value fruit and vegetable crops. The tolerance for damage is very low, especially for fruits intended for export markets. Quarantines are placed on commodities that are likely to be infested with tropical fruit flies. Fruits may be treated in various ways to kill the immature stages and thereby minimize the risk that the pests will spread via the marketing channels. However, the treatments are costly and may affect the quality of the products. Also, restrictions or bans on certain chemical treatments have made the certification of commodities for export difficult.

Because of the low damage tolerance, there is worldwide interest for acceptable suppressive measures that will maintain populations at very low levels or that can be employed for eradication programs. Ordinary pest-managment procedures generally do not provide the degree of control that is required to adequately cope with tropical fruit fly problems. Insecticide sprays have been developed for eradication or control (Steiner et al. 1961). However, they can cause harmful ecological imbalances and, when used in residential areas, can arouse great public concern over their potential to harm humans and animals. The sterile insect technique is being used with considerable success in certain programs (International Atomic Energy Agency 1982). This technique alone is not practical, however, when the population density of the target pest is high. While insecticide sprays can be used to suppress populations to the extent that they can subsequently be eliminated or managed by sterile flies, use of this integrated approach to attack fruit fly populations has been limited because of the aforementioned opposition to insecticide sprays. Insect attractants, methyl eugenol in particular, offer methods of eliminating or suppressing certain species (Steiner et al. 1965, 1970; Koyama et al. 1984). Generally, however, insecticides must be added to the attractants to kill the flies.

For these reasons, I gave special attention to appraising the potential of using parasite releases to suppress tropical fruit fly popula-

Figure 4. *Biosteres tryoni.* This insect is one of the most effective of the species that parasitize the Mediterranean fruit fly in Hawaii. Tropical fruit flies are the most destructive insect pests of fruit in many parts of the world. Although the fruit flies are parasitized at high rates by a number of parasite species, the tolerance for damage to high-value fruit is so low that natural parasitism does not provide adequate control. Because of the low tolerance for damage and the constant risk that a number of fruit fly species will spread to sensitive areas through the shipment of infested fruits, there is worldwide interest in measures that can be used to eradicate fruit fly populations or to reduce them to near-extinction levels. These pests can be eradicated or suppressed to very low levels by the use of chemical insecticides, but this method is not acceptable to the public because of potential health and environmental hazards. The use of sterile flies is an acceptable alternative, but this technique is not practical against moderate to high natural populations.

Theoretical estimates of the influence of parasite releases in tephritid fruit fly ecosystems indicate that natural populations of several fruit fly species could be reduced to very low levels by the parasite augmentation technique. Therefore, the integration of parasite and sterile fly releases may prove to be an effective, practical, and ecologically acceptable means of solving many of the tropical fruit fly problems in various parts of the world. (Photograph, courtesy of USDA Tropical Fruit and Vegetable Research Laboratory, Honolulu, HI.)

tions to levels that can then be eliminated or managed by the sterile insect technique. Fortunately, nature has evolved a number of parasites that are primary parasites of tropical fruit fly species (Bess et al. 1961, Clausen et al. 1965). Tropical fruit flies can be colonized and mass-produced at low cost. Therefore, the prospects are excellent that a number of candidate parasites could also be mass-produced and at reasonable costs for use in parasite augmentation programs.

Little is known, however, about the actual numbers of parasites that exist in natural populations and the numbers of parasites that will have to be released to achieve rates of parasitism that would suppress fruit fly populations to tolerable levels. No field trials have been made to determine the effectiveness of the augmentation technique against isolated fruit fly populations. Indirect, deductive procedures to be described have therefore been used to appraise the potential of this approach for suppressing tropical fruit fly populations. The following fruit fly species are included in this appraisal: Mediterranean fruit fly, *Ceratitis capitata* (Wied); oriental fruit fly, *Dacus dorsalis* (Hendel); and the Caribbean fruit fly, *Anastrepha suspensa* (Loew). *Biosteres tryoni* (Cameron), *B. arisanus* (Fullaway), *B. longicaudatus* (Ashmead), and *B. vandenboschi* (Fullaway) are the parasite species evaluated.

Suppression of Medfly Populations

The medfly is the most important of the tropical fruit fly species. It exists in tropical and subtropical areas under a wide range of ecological conditions. The most practical and effective parasites for augmentation purposes are likely to vary depending on the environment, the prevailing host fruit, the degree of host continuity, and other factors. Most of the parasites of tropical fruit flies are solitary species that develop in the host larvae.

In this analysis, *Biosteres tryoni*, considered by Wong and Ramadan (1987) to be an excellent medfly parasite, will be evaluated (fig. 4). It has a strong affinity for the medfly. When originally introduced into Hawaii, it became the dominant parasite, causing parasitism up to 50 percent (Pemberton and Willard 1918). After the establishment of the oriental fruit fly and the associated egg-larval parasite, *B. arisanus*, interspecific competition considerably affected the role of *B. tryoni* as a natural control agent. Nevertheless, its basic host-parasitization capability remains unaltered. *B. tryoni* has a preference for third-instar larvae and has a fecundity of about 90 eggs.

Natural Coexistence Pattern of Medfly and *Biosteres tryoni* Populations in Hawaii

Coexistence Model

The procedure used in this study is to develop a population model that reflects actual and relative numbers of parasites and hosts that coexist in natural populations and how the two populations are likely to grow from the normal low-density to the normal high-density levels. Populations of the medfly vary greatly in different ecosystems, and the normal maximum density in some areas may exceed the normal minimum by 100-fold. However, under conditions in Hawaii, the range in the normal density is more likely to be about 10-fold (Vargas et al. 1983, Wong et al. 1984). The fact that three to five generations may be produced during the year suggests that even under favorable conditions for reproduction, the average rate of increase per cycle is not likely to exceed 2- to 3-fold. Apparently, *B. tryoni is* largely dependent on the medfly for reproduction and survival in Hawaii. For this analysis, the normal rate of parasitism due to *B. tryoni* is assumed to average about 25 percent. This rate is slightly lower than the average rate observed by Wong and Ramadan (1987). However, the average number of hosts parasitized per female would be essentially the same whether the normal average rates of parasitism were higher or lower.

After developing a number of preliminary models by using different values for the various parameters, I selected the model shown in table 8 as representative of the coexistence pattern of medfly and *B. tryoni* populations in Hawaii. It is based largely on the theories relating to parasite-host relationships discussed in chapter 2 and the following basic assumptions:

1. The medfly generation of lowest population consists of an average of 10,000 males and 10,000 females per square mile of habitat.

2. Medfly populations in Hawaii increase by 2- to 3-fold per cycle during periods favorable for reproduction. As the populations grow, density-dependent suppression forces tend to slow down the rate of growth. The population shown increases by about 12-fold during a favorable season and then begins to decline to its normal low level.

3. The ratio of the number of adult *B. tryoni* to the number of adult medflies in typical populations is about 1:3. This ratio is based on the assumptions that the average parasitization rate is 25

Table 8
Basic model depicting the estimated coexisting pattern of medfly and *Biosteres tryoni* populations in a natural environment (see text for more details)

Parameter[1] (1 square mile)	Generation		
	1	2	3
Medfly females	10,000	30,000	72,900
Host larvae produced per female (3d instar)	40	36	30
Total host larvae	400,000	1,080,000	2,187,000
B. tryoni females	3,333	10,000	24,300
Ratio, adult parasites to adult hosts	1:3	1:3	1:3
Host larvae parasitized per female	30	27	22
Total host larvae parasitized	100,000	270,000	534,600
Parasitism, percent	25	25	25
Host larvae not parasitized	300,000	810,000	1,652,400
Survival of parasites and hosts to adult stage, percent	20	18	15
Adult medflies, next generation	60,000	145,800	247,860
Adult *B. tryoni*, next generation	20,000	48,600	80,190
Increase rate of the host and parasite populations	3.0	2.4	1.7

[1]In view of the increasing host and parasite densities, the average number of host larvae produced per female medfly, the average number of host larvae parasitized per female *B. tryoni*, and the survival rates of the parasite and host larvae are assumed to decline. Therefore, the rates of increase of the populations tend to diminish, but host fruit availability and other factors could alter the population trends.

percent and that the survival rates of the immature parasite larvae and the unparasitized host larvae to the adult stage tend to be equal. The *actual* survival rates may be higher or lower, but the *relative* survival rates remain essentially the same.

4. The proportion of the host population parasitized tends to remain constant at 25 percent, irrespective of the host density. Therefore, if the survival rates are equal, the ratio of adult parasites to adult hosts will also tend to remain constant at 1:3.

5. The number of host larvae (3d instar) produced by the medfly female during its lifetime is estimated to average 40 during the most favorable periods for reproduction. The average is assumed to decline as the host density increases.

6. The average number of host larvae that the *B. tryoni* female parasitizes during its lifetime is assumed to be 30 under the conditions allowing the medfly female to produce an average of

40 hosts. The average decreases as the host and parasite population increases.

Discussion of the Model

The rationale for the values assigned to the various parameters will be discussed in some detail. A similar rationale was used in the development of models depicting the coexistence patterns for the other fruit fly and associated parasite species included in this investigation.

The most important parameters for the model are (1) the actual and relative numbers of hosts and parasites in natural populations, (2) the actual and relative average numbers of host larvae produced by the medfly females during their lifetime and the actual and relative numbers of host larvae parasitized by the associated *B. tryoni* females during their lifetime, and (3) the actual and relative survival rates of the immature parasites and unparasitized larvae to the respective adult stages. Appropriate values for these parameters each cycle will reflect the population trends for the two organisms from the normal low density to the normal high density and the rates of parasitism that will occur. A key parameter is the estimated average host-parasitization capability of the *B. tryoni* females. This value will be used, with some modification, together with a given ratio of adult parasites to adult medflies in the host ecosystem to extrapolate the proportion of the medfly population that will be parasitized. Quantitative information on the various parameters is either limited or totally lacking. Therefore, indirect, deductive procedures were used to quantify the parameters. Several theories on the factors that govern the parasitization process in natural environments, as discussed in chapter 2, were considered in the quantification.

Like other models developed for the various parasite-host associations, the model shown in table 8 was selected from among several medfly-*B. tryoni* models that had also been developed and analyzed. From a practical standpoint, the most important parameter is the accuracy of the estimated number of adult medflies that normally exist per square mile of medfly habitat during the generation of lowest density. The number of insects required for the desired parasite-to-host ratio or the desired sterile-to-fertile ratio hinges on the accuracy of the estimate. While actual numbers of medflies that exist per unit area during low periods may exceed or fall below the estimate of 10,000 of each sex per square mile, it seems unlikely that the actual numbers would differ from the estimate by a wide margin. Thus, the model should serve its purpose. I am confident that the estimated relative values for the other parameters will deviate very little from true values. The rationale for the various estimates will be discussed in some detail.

The model depicts a seasonal increase of about 12-fold for a typical medfly population in Hawaii. The rate of increase is assumed to range from 1.7- to 3.0-fold per cycle, with an average of 2.3-fold per cycle. If the average is assumed to be 25 percent higher, or about 3.1-fold per cycle, the population would grow 30-fold in three cycles. Trap capture data would not support such a high rate of increase. If the average is assumed to be 25 percent lower, or about 1.7-fold, the peak population would exceed the low population by only about 5-fold. Therefore, the estimated typical rates of increase per cycle and for the season as depicted in the model would seem reasonable for a stabilized medfly population. In favorable new areas of spread, the average rate of increase per cycle is likely to be much higher.

If the model is reasonably representative of the number and dynamics of medfly populations, the number and dynamics of an associated *B. tryoni* population can be estimated on the basis of certain theories that have been discussed earlier in this report. One of the most important is the theory that the survival rates of the immature parasites and unparasitized host larvae to the respective adult stages are essentially equal at all host and parasite densities. Since the normal rate of parasitism due to *B. tryoni* is assumed to be about 25 percent, the normal ratio of adult *B. tryoni* to adult medflies should be on the order of 1:3 (25:75). If average parasitism exceeds or falls below the assumed 25 percent, the ratio would differ, but the ratio of *B. tryoni* adults to medfly adults can be estimated on the basis of the normal rate of parasitism and the comparable survival theory. Also, if a given ratio of adult parasites to adult hosts is known to cause a certain rate of parasitism, the rate of parasitism to expect from increasing ratios can be calculated, as previously described in chapter 2.

As noted earlier, one of the most important parameters is the average number of hosts that the females in a parasite population will normally parasitize during their lifetime. Another important parameter is the average number of hosts of the stage parasitized that is normally produced by females of the host population. The medfly female is estimated to produce an average of 40 large host larvae during generation 1, and the average is assumed to decline as the medfly population grows. The associated *B. tryoni* female is estimated to parasitize an average of 30 host larvae during generation 1, and the average is assumed to decline as the parasite and host populations grow. However, the relative average numbers of host larvae produced and host larvae parasitized are assumed to remain reasonably constant. A constant relationship between the average rates of survival of the immature parasites and unparasitized host larvae to the respective adult stages helps maintain a constant rate of parasitism, irrespective of the host density. A rather constant rate of parasitism seems to be characteristic for many parasite-host associations, as discussed in chapter 2.

The estimate that the normal average number of large larvae produced by the medfly female ranges from 30 to 40 may seem low. The basis of the estimate is the high mortality of the medfly. According to Vargas et al. (1984), the medfly female has the capability of depositing about 1,000 eggs. Yet, as discussed earlier, the average rate of increase in medfly populations is likely to range between 2- and 3-fold per cycle. An increase rate of 3-fold during generation 1 would mean an average of only six adult progeny produced per female. Thus, the cumulative mortality rate would be about 99.4 percent, or a survival rate of only 0.6 percent. I estimate that 96 percent of the mortality occurs before the larvae reach the third instar and that about 20 percent of the third-instar larvae normally survive to the adult stage. The average number of larvae produced may be higher or lower and the normal survival rates may be correspondingly lower or higher, but the key parameter is the average rate of increase in the medfly population.

The rationale for the estimated average number of host larvae parasitized and for the survival rates of the immature parasites and unparasitized host larvae to the adult stage is likewise based on deductions. While the estimated *actual* values for the parameters may not be highly accurate, I believe that the *relative* values are very accurate. This belief is supported by the fact that the population trends in the model will change drastically if even rather minor changes are made in the estimated relative numbers of hosts produced and hosts parasitized and the relative survival rates of the parasites and hosts. For example, if the female parasites are assumed to parasitize as many hosts as are produced by the medfly females and the survival rates of the parasites and hosts are assumed to be equal, or if the rate of survival of the parasite larvae is 25 percent higher than the rate of survival of the unparasitized host larvae, the parasite and host populations would become grossly imbalanced within four to five cycles. Parasitism would increase to a level of 75–80 percent, and the survival of the medfly population would be jeopardized. Survival of the *B. tryoni* population would then also be in jeopardy. If the *B. tryoni* females are assumed to parasitize only half as many host larvae as are produced by the medfly females, the ratio of *B. tryoni* to medfly adults would be reduced to such a low level that the rate of parasitism would approach zero within four to five cycles, and the survival of the parasite population would be in doubt.

The model presented does not account for a number of variables that will influence the size and trends of medfly populations. The variables include the amount and distribution of host fruit and variations in weather conditions. The abundance of medflies will vary from season to season and habitat to habitat. But *B. tryoni* is a highly specialized and highly efficient organism that has proven its ability to parasitize enough hosts to maintain viable populations under a wide range of conditions. In the section to follow, we will consider what the probable rates of parasitism will be and how the dynamics of medfly populations will be influenced when *B. tryoni* adults are released in sufficient numbers to greatly alter the normal numerical relationship. The values assigned for the key parameters that govern the normal numerical relationships in natural populations will be used, with some modification, to estimate the influence of such releases.

Influence of *Biosteres tryoni* Releases

Suppression Model

Table 9 shows the theoretical influence of *B. tryoni* releases on the rates of parasitism and on the dynamics of a medfly population of the nature shown in the basic model, table 8. Releases will be made at the rate of 56,000 parasites of both sexes per square mile during generation 1, when the natural population is assumed to average 20,000 medflies of both sexes. The releases would lead to a 2.8:1 ratio of adult parasites to adult hosts. The theoretical effects of the releases during the generation that the releases are made and in subsequent cycles are based on the parameter values assigned to natural coexisting populations and the various theories discussed in chapter 2. The values assigned to the different parameters and how the calculations were made are largely self-explanatory.

Discussion of Results

In view of the abnormally high parasite population, which would undoubtedly lead to above-normal predation, the host-parasitiza-

Table 9
Suppression model depicting the estimated influence of *Biosteres tryoni* releases in medfly ecosystems (see text for details)

Parameter (1 square mile)	Generation			
	1	2	3	[1]4
Medfly females	10,000	7,400	3,231	460
Host larvae per female[2]	40	42	44	50
Total host larvae	400,000	310,800	142,164	23,000
B. tryoni females	28,000	32,600	29,403	16,599
Ratio, adult parasite to adult hosts	2.8:1	4.4:1	9.1:1	36:1
Host larvae parasitized per female[3]	24	22	18	—
Total parasite-host encounters	672,000	717,200	529,254	—
Ratio, parasite-host encounters to host larvae present[4]	1.7:1	2.3:1	3.7:1	—
Parasitism percent	81.5	90.1	97.3	—
Host larvae parasitized	326,000	280,030	138,326	—
Host larvae not parasitized	74,000	30,770	3,838	—
Survival to adults, percent	20	21	24	—
Medflies, next generation	14,800	6,462	921	—
B. tryoni, next genertion	65,200	58,806	33,198	—

[1]Dashes indicate that the data are not estimated.
[2]The average number of host larvae produced per female medfly is assumed to increase as estimated because of the decrease in host density.
[3]The average number of host larvae parasitized per female parasite is assumed to decrease as indicated because of an increasing parasite population and declining host density.
[4]See table 1 to calculate the rates of parasitism and escape from parasitization by using the ratio of parasite-host encounters to hosts present.

tion capability of the female parasite is assumed to average 24. Thus, the adjustment factor for this key parameter is 20 percent. However, because of the large number of parasites present, enough of the host larvae would be parasitized to result in 81.5 percent parasitism during the first generation. The percentage allows for the encounter and reparasitization of previously parasitized larvae. Since the ratio of parasites to hosts would be about eight times the ratio assumed to cause 25 percent parasitism, the estimate of 81.5 percent parasitism would seem reasonable. Based on the 3-fold potential rate of increase for a natural population during generation 1, the medfly population would decline by 26 percent to 7,400 medfly females per square mile during generation 2. Since the rates of survival of the immature parasites and unparasitized host larvae to the respective adult stages are assumed to be equal, the ratio of adult parasites to adult hosts during generation 2 would be 81.5:18.5, or 4.4:1. This ratio is substantially greater than the 2.8:1 ratio resulting from the parasite releases during generation 1.

The estimated influence of the parasites during generation 2 is based on relatively minor adjustments in the average number of host larvae produced by the medfly females and the average number parasitized by the female parasites. In view of the lower adult host population, the average number of larvae it produces is assumed to increase by about 5 percent. However, the parasite population has theoretically increased by about 16 percent. Therefore, the average number of host larvae parasitized by the female parasites is assumed to decrease by about 8 percent. The estimated 22-percent decline in the larval host population is assumed to have little or no effect on the host-finding capability of the female *B. tryoni*. The net effect of these changes in parameter values would be an increase in the ratio of parasite-host encounters to hosts present and an increase in parasitism to 90.1 percent. As a result, a decline in the medfly population to about 68 percent below normal would be expected by generation 3. Thus, the ratio of adult parasites to adult hosts would theoretically increase by more than 3-fold by generation 3.

The estimated influence during generation 3 is based on additional adjustments in parameter values. The average number of host larvae produced per female medfly is assumed to increase by about 10 percent above normal because of the lower host density. The estimated average number of hosts parasitized by the *B. tryoni* females is assumed to decrease by about 25 percent below the average during generation 1. It is questionable as to whether such a decrease in the average host-parasitization capability of the *B. tryoni* females is justified. But theoretically, the large increase in the parasite-to-host ratio would increase the rate of parasitism to 97.3 percent during generation 3.

Although considerable adjustments were made in parameter values that favor the medfly population, the population by generation 4 is estimated to decline to about 95 percent below the normal low density. On the other hand, the *B. tryoni* population would decline less than 50 percent. Thus, the ratio of parasites to hosts would steadily increase to 36:1 during generation 4. No estimate is made of the influence of the parasites during generation 4. The influence would depend on the average host-parasitization capability of the females when the host population is this low. As discussed in chapter 2, the highly developed host-guidance mechanism possessed by parasites may permit them to find enough hosts to maintain a viable population even though the host population is near extinction. If normal survival of the parasites in the parasitized hosts is 25 percent under favorable conditions, the females would have to parasitize an average of eight host larvae to produce enough progeny to equal the parent population. At the ratio indicated during generation 4, an average of eight parasitized larvae per female would result in approximately 99 percent parasitism. If so, the parasite releases alone would almost eliminate the pest population.

In general, there is reason to believe that the suppression model reasonably depicts the results that could be achieved if *B. tryoni* is released in the numbers proposed. Of particular significance is the progressive increase in parasitism until the host population is reduced to a very low level. As discussed earlier, such an increase would be new to science and would open up the possibility of applying highly effective strategies for regulating pest populations. It would seem important to undertake appropriate field-release experiments to determine whether such an increase will occur.

The possibility that a parasite population developing on its own will lead to virtual elimination of a host population if the initial release rate is high enough may seem beyond realism. But in theory, we would create numerical relationships between parasite and host populations that could never be approached in natural populations, and such distorted numerical relationships have never been created artificially throughout a pest-host ecosystem. We might view the model from a broad perspective to consider the possibility that it does indicate the potential of the augmentation technique. The data presented for generations 1 and 2 would seem plausible, considering that although the normal numerical relationship between the parasite and host populations is likely to be about 1:3, the ratio was increased artifically by about 8-fold and then naturally, while the medfly population was still quite high, to about 13-fold. If a ratio of 1:3 can be expected to cause 25 percent parasitism, it would seem plausible to expect about 80 percent parasitism when the ratio is 8 times higher and about 90 percent when the ratio is 13 times higher. By generation 3, the increase in ratio would be about 27 times above normal; and by generation 4, more than 100 times higher than could be expected in a natural population. In theory, we are dealing with relative numbers that deviate from normal numbers to such a degree that we have no way of relating the theoretical results to the observed results of natural parasitism.

As a minimum, the model would seem to indicate that the prospects are highly favorable that the release of parasites can provide an effective and practical means of reducing medfly populations to a level that could subsequently be eliminated by releasing relatively few sterile flies. As emphasized at the beginning of the chapter, there is an urgent need for ecologically acceptable measures to greatly suppress or eradicate fruit fly populations.

Scenario of a Medfly Eradication Program Based on Parasite and Sterile Fly Releases and an Estimate of the Costs

The theoretical appraisal of the influence of *B. tryoni* releases in medfly ecosystems was focused on the numbers of parasites that would have to be released to achieve a high degree of host suppression. However, the numbers have no practical significance, no matter how favorable the results might be, unless they are related to costs likely to be involved in rearing and releasing the insects. To accurately estimate such costs before the technology has been thoroughly investigated is difficult. Nevertheless, I feel that scientists have made enough progress in insect rearing technology that meaningful estimates can be made of eventual costs of rearing various kinds of insects—particularly, the various species of tropical fruit flies.

Estimating the degree of control and costs in operational programs from the results of theoretical suppression models is obviously difficult, and the accuracy will be uncertain. However, based on the highly encouraging results that have been discussed, I will describe a hypothetical program and the magnitude of the costs likely to be involved.

The medfly habitats in Hawaii have been estimated to total about 1,000 square miles[1]. The area where year-round reproduction occurs may be substantially less. While the number of medflies per unit area will vary greatly, it is assumed, as noted, to average 20,000 per square mile during the generation of lowest density. The theoretical suppression model indicates discrete generations, as might be expected for insects that have overwintering periods. However, in tropical areas, all of the life stages will be present at any given time. There will also be considerable overlapping of generations, and some adults will likely survive for two generations or longer. Such variables were not considered in the model.

While the theoretical model in table 9 indicates that parasite releases for only one generation can lead to near elimination of the

[1]U.S. Department of Agriculture, Animal and Plant Health Inspection Service. 1984. Eradication of the tri-fly complex from the State of Hawaii. Draft Environmental Impact Statement (unpublished), 330 pp.

medfly population, a more conservative assumption is that releases will be required for several generational cycles. It is therefore proposed that an average of 60,000 *B. tryoni* of both sexes be released per square mile for three successive cycles. The releases would begin when conditions are favorable for reproduction and survival of the hosts and parasites. To the degree practical, releases would be made proportional to the densities of the medflies in the various habitats. Thorough surveys prior to and during the release periods would therefore be vital for such a program.

Theoretically, releasing parasites in generations 1–3 should lead to the near elimination of the medfly population by generation 4. A more conservative assumption, however, is that by generation 4, the population will have been reduced by only 95 percent and will therefore still consist of 1,000 medflies of both sexes per square mile. Also, disregarding the probability that the parasite progeny would continue to provide considerable control, we will consider the influence of sterile fly releases only. The release of 100,000 sterile flies of both sexes per square mile would theoretically result in an overall sterile-to-fertile ratio of 100:1. If the released flies are 25 percent competitive, the effective ratio would be 25:1. Even if the potential increase rate of the native flies at such a low density were as high as 10-fold, a 25:1 ratio would mean that the natural population should be reduced to 400 per square mile. The continued release of 100,000 sterile flies per square mile would, in such event, result in an overall sterile-to-fertile ratio of 250:1 and an effective ratio of about 62:1 during generation 2. The sterile-to-fertile ratio by the third generation after releases began should be several thousand to one, and the probability of fertile matings would be nil. Some sterile fly releases should continue for several cycles, however, for security reasons.

Based on this scenario, which may be conservative in several respects, a total of 180,000 *B. tryoni* and about 400,000 sterile medflies would be required per square mile of medfly habitats. While mass-rearing methods for *B. tryoni* have not been perfected, considerable progress has already been made in the U.S. Department of Agriculture's investigations in Hawaii (Wong et al. 1991). For this gross estimate, the eventual cost of rearing and releasing *B. tryoni* is estimated to be $2,000 per million. The cost of rearing and releasing the sterile medflies is estimated at $500 per million. Thus, the cost of the insect components of such a program would be about $560 per square mile. For 1,000 square miles, the cost would only be on the order of $560,000. The major costs for such a program would doubtless be for survey activities, general administration, and supervision. However, excluding costs of the rearing facilities, it may be possible to eradicate the medfly from Hawaii by the procedure proposed at a cost of $1 million or $2 million.

For comparative purposes, the cost of a single insecticide application is likely to be about $1,000 per square mile, and at least eight insecticide applications would doubtless be required for eradication. Thus, on a per-square-mile basis, the costs would be about $8,000 for the chemical spray approach and $560 for the integrated release of parasites and sterile flies.

Releasing the parasites and sterile flies according to the scenario described would not, however, be the most effective procedure to follow. Chapter 3 presents the theory of integrating parasite and sterile insect releases to greatest advantage. Accordingly, by far the most effective and practical procedure would be to release parasites and sterile flies concurrently beginning with the first generation. We might assume the release of 100,000 sterile flies and 60,000 parasites during generation 1.

If the sterile flies inhibit reproduction of the normal flies by only 50 percent, the effective ratio of adult parasites to adult reproducing medflies would be about 6:1 rather than 3:1. The rate of parasitism would then be about 98 percent. This rate, in turn, would increase the sterile-to-fertile ratio by about 10-fold during generation 2. In turn, the ratio of adult parasites to reproducing adult hosts during generation 2 would theoretically be about several hundred to one. The potential of this integrated approach to medfly suppression is difficult to envision, and certainly it cannot be fully realized unless the approach is directed at total pest populations. However, the basic principles of suppression seem biologically sound.

Suppression of Oriental Fruit Fly Populations

When the oriental fruit fly, *Dacus dorsalis*, became established in Hawaii in about 1946, its population increased greatly because of the favorable climate and abundance of fruit. A major effort was made by cooperating agencies to control the pest by the introduction of biological agents (Bess et al. 1961, Clausen et al. 1965). As a result, several species of parasites became well established and eventually reduced the steady density of the pest population by an estimated 94 percent (Newell and Haramoto 1968). Three species were primarily responsible for the suppression, namely, *Biosteres longicaudatus*, *B. vandenboschi*, and *B. arisanus*. *B. arisanus* became the dominant species, achieving a rate of larval parasitism as high as 75 percent (Newell and Haramoto 1968). Although interspecific competition caused a decline in the rates of parasitism achieved by the other two species, together they initially accounted for a parasitization rate of about 45 percent (Vanden Bosch and Haramoto 1953). The actual rate may have been higher, since larvae of all ages were included in the collections. *B. arisanus* is an egg-larval parasite, whereas *B. longicaudatus* is a larval-pupal parasite and *B. vandenboschi* a larval parasite. Normally, therefore, these last two species would be expected to successfully parasitize only those host larvae that have escaped parasitism by *B. arisanus.* Nevertheless, *B. longicaudatus* and *B. vandenboschi* are inherently capable of achieving high rates of parasitism and seem well adapted to environmental conditions in Hawaii. Therefore, they should be regarded as promising candidates for augmentation purposes. Indeed, as will be shown, they may be more effective and practical than *B. arisanus* for augmentation purposes.

Even though the introduced parasites greatly reduced the oriental fruit fly populations in Hawaii, they eventually reached stability with the environment and with the host at a level that permits the oriental fruit fly to be a continuing threat to Hawaiian agriculture and also a continuing threat to be inadvertently transported to the U.S. mainland and other areas. Because the oriental fruit fly is frequently introduced into the continental United States and because the costs of eradicating infestations are high, there is interest in eradicating the oriental fruit fly, as well as the medfly, from Hawaii.

Isolated populations of the oriental fruit fly can be eradicated by a male annihilation technique that involves using a formulation containing methyl eugenol (a powerful attractant) and insecticides to kill the attracted male flies (Steiner et al. 1970, Koyama et al. 1984). The sterile insect technique is also effective for eradication but only of low-level populations (Steiner et al. 1970). Thus, when the pest populations are moderate to high, insecticide sprays or methyl eugenol formulations containing insecticides must be used to greatly reduce the populations before sterile flies can be used for eradication.

Since the parasite augmentation and the sterility techniques are target-pest specific and create no health hazards or significant environmental risks, the use of these techniques, alone or appropriately integrated, should be entirely acceptable to the public. The potential for using the augmentation technique in Hawaii for oriental fruit fly suppression will be analyzed for the three parasite species named.

Characteristics of *Biosteres arisanus*

The biology, ecology, and behavior of *B. arisanus* and its relationship to the oriental fruit fly and, to some degree, the medfly, have been thoroughly investigated by a number of scientists for many years. *B. arisanus* parasitizes the host eggs but completes development in the larval stages of the hosts. One host yields one adult parasite. Newell and Haramoto (1968) made an indepth study of the relationship between *B. arisanus* and the oriental fruit fly. The information they obtained supports several of the key hypotheses and assumptions I made in appraising the potential of *B. arisanus* and other parasites to suppress pest-host populations.

As already noted, parasitism data showed that the rate of larval parasitism due to *B. arisanus* after its establishment was as high as about 75 percent. The rate at which *B. arisanus* parasitized host eggs was much higher, however.

In conducting their investigation, Newell and Haramoto frequently observed very high rates of parasitism in eggs—up to 100 percent, on occasion. They estimated that when larval parasitism averaged 75 percent, egg parasitism will have averaged about 95 percent and the ratio of adult parasites to adult hosts will be about 3:1. These observations suggest that if the ratio of adult parasites to adult hosts is high enough, the rate of egg parasitism may approach 100 percent. The likelihood of near-100-percent egg parasitism in natural populations is very important, because we will be considering theoretical results of ratios that exceed the ratios in natural populations by 10-fold or more.

Newell and Haramoto made several other observations that helped me assign values to key parameters in the population models. One of the most significant observations was that parasitized eggs suffered high mortality due to trauma and pathogenic organisms. The authors reported that about 80 percent of the parasitized eggs were killed when the egg parasitization rate was 95 percent. Egg mortality increased or decreased, depending on the percentage of eggs that were parasitized. The rate of superparasitism no doubt determines the egg mortality. Although not mentioned by the authors, the high mortality of parasitized host eggs can be shown to be an expression of a natural regulating mechanism for limiting the ratio of adult parasites to adult hosts that can coexist. The rate of survival of the immature parasites to the adult stage and the rate of survival of the large unparasitized host larvae to the adult stage are assumed to be approximately equal. Thus, if larval parasitism averages 75 percent when the egg parasitism had been 95 percent, the ratio of adult parasites to adult hosts will be near 3:1, instead of 19:1.

Particularly significant were the observations by Bess et al. (1963) and Haramoto and Bess (1970) that the average rate of parasitism due to *B. arisanus* tends to remain constant throughout the year despite the usual fluctuations in the host population, abundance of fruit, and weather conditions. This constancy is important evidence supporting the observation that among many parasite-host complexes, host density and rates of parasitism are not positively correlated. Parasitism eventually stabilized at about 65 percent. In the meantime, the steady density of the oriental fruit fly population declined by about 94 percent.

It might be argued that the decline in rate of parasitism from a high of about 75 percent to about 65 percent was due to the declining steady density of the host population. If so, some positive correlation between the host density and rate of parasitism would have been noted, but none was. A more plausible explanation may be that the parasite population merely established a lower steady density in response to the overall environmental influences.

Natural Coexistence Pattern of Oriental Fruit Fly and *Biosteres arisanus* Populations in Hawaii

Since *B. arisanus* proved to be the most effective parasite introduced into Hawaii to biologically control the oriental fruit fly, it would seem to be the species of choice for augmentation purposes. Therefore, its role as a natural parasite was considered first. The model in table 10 shows the average numbers of *B. arisanus* and *D. dorsalis* estimated to coexist in natural environments in Hawaii. As noted before, information on actual numbers of insects in natural populations is usually lacking. I estimate, however, that an average of 40,000 oriental fruit fly adults of both sexes are present per square mile when the populations are normally low. Actual numbers per square mile are likely to be much higher or lower in many habitats, but even so, this average for about 3,000 square miles of oriental fruit fly habitats in Hawaii may be a conservative estimate. The normal maximum population is estimated to exceed the normal minimum by about 5- to 10-fold.

For this analysis, the natural rate of larval parasitism is assumed to average 60 percent, a rate that is slightly lower than the average reported by Newell and Haramoto. Based on the comparable

Table 10
Basic model depicting the estimated coexistence pattern of *Dacus dorsalis* and *Biosteres arisanus* populations in Hawaii

Parameter (1 square mile)	Generation 1	
	1	[1]2
D. dorsalis females	20,000	48,600
Eggs per female	400	360
Total host eggs	8,000,000	17,496,000
B. arisanus females	35,000	83,830
Ratio, parasites to fruit flies	1.75:1	1.75:1
Eggs parasitized per female	300	270
Total parasite-host (-egg) encounters	10,500,000	22,634,100
Ratio, parasite-host encounters to eggs present	1.3:1	1.3:1
Eggs parasitized, percent[2]	73	73
Eggs parasitized, number	5,840,000	12,772,080
Eggs not parasitized, number	2,160,000	4,723,920
Parasitized eggs killed by trauma, percent[3]	45	45
Eggs successfully parasitized, number	3,212,000	7,024,644
Survival parasitized and unparasitized eggs to large larvae, percent	20	18
Parasitized large larvae, number	642,400	1,264,436
Unparasitized large larvae, number	432,000	850,306
Large larvae parasitized, percent	60	60
Survival of large larvae to adults, percent	22.5	20
Adult fruit flies, next generation	97,200	170,061
Adult *B. arisanus*, next generation	144,542	252,887
Adult *B. arisanus* females next generation[4]	83,830	147,427
Increase rates of fruit fly and parasite	2.4	1.8

[1]In view of the higher host and parasite densities, the average number of eggs produced per female host, the average number of eggs parasitized per female, and the survival rates are assumed to be reduced by 10 percent. These reductions result in lower increase rates for the host and parasite populations.
[2]See data in table 1 to calculate the rates of parasitism and escape from parasitism by using the ratio of parasite-host encounters to hosts present.
[3]This estimate is considered consistent with observations made by Newell and Haramoto (1968).
[4]The ratio of females to males is assumed to be 1.4:1

survival theory, the ratio of adult *B. arisanus* to adult *D. dorsalis* will normally be about 60:40 or 1.5:1. Wong et al. (1984) noted that the sex ratio of *B. arisanus* generally favors the females by a ratio of about 1.4:1. Therefore, if the ratio of both sexes of the parasites to the hosts is 1.5:1, the ratio of female parasites to female hosts is more likely to be near 1.75:1. It is assumed, therefore, that 35,000 female *B. arisanus* will coexist with 20,000 female *D. dorsalis*.

Since *B. arisanus* females parasitize host eggs, estimates of the average number of host eggs produced by *D. dorsalis* females and the average number parasitized by the *B. arisanus* females are necessary. After considering a number of values for these two important parameters, I concluded that 400 eggs per female fly and 300 parasitized host eggs per female parasite would be realistic averages. Vargas et al. (1984) determined that the *D. dorsalis* female in laboratory colonies deposits an average of about 1,400 eggs. The mortality in nature during all of the life stages of insects is known to be high. The assumption is made that *D. dorsalis* populations have an intrinsic rate of increase ranging from 2- to 3-fold during periods favorable for reproduction and survival. This assumption is partly based on data (Bess et al. 1963, Haramoto and Bess 1970) indicating that the maximum and minimum *D. dorsalis* populations are not likely to fluctuate by more than 5- to 10-fold during the season. The model shows an increase rate of 2.4-fold during cycle 1 and 1.8-fold during cycle 2. The survival rate would thus be about 0.3 percent even when conditions for reproduction are favorable.

The female *B. arisanus* is assumed to parasitize an average of 300 host eggs during its lifetime under conditions assumed for generation 1. Based on these values, 73 percent of the host eggs would be parasitized during generation 1. The allowance made for the mortality of parasitized eggs is considered consistent with observations made by Newell and Haramoto (1968). The survival rates of the immature parasites and unparasitized host larvae are assumed to be equal. The populations of both *D. dorsalis* and *B. arisanus*, as shown in the model, would grow at a similar rate. Both species would reproduce successfully, and the rate of parasitism would remain constant.

In general, the parameter values for the basic model seem consistent in every respect with data and observations reported by the various authors cited. However, the reliability of the estimated numbers of oriental fruit flies and parasites that coexist might be questionable because of the paucity of relevant field data. The estimate of actual numbers and their ratio are highly important because they determine the actual numbers of the parasites that must be released to achieve various rates of parasitism and host suppression.

Influence of *Biosteres arisanus* Releases

Suppression Model

Table 11 is a model showing the estimated suppressive influence of *B. arisanus* releases in ecosystems having a *D. dorsalis* population of

Table 11
Suppression model depicting the estimated influence of *Biosteres arisanus* releases in *Dacus dorsalis* ecosystems (see text for details)

Parameter[1] (1 square mile)	Generation		
	1	2	3
Female *D. dorsalis*	20,000	9,000	9,017
Eggs deposited per female	400	440	440
Total eggs deposited	8,000,000	3,960,000	3,967,480
B. arisanus females	100,000	31,482	20,558
Ratio, adult parasites to adult hosts	5:1	3.5:1	2.3:1
Eggs parasitized per female	240	300	310
Total parasite-host (-egg) encounters	24,000,000	9,444,600	6,372,980
Ratio, parasite-host encounters to host eggs present[2]	3:1	2.4:1	1.6:1
Host eggs parasitized, percent	95	91	80
Host eggs parasitized, number	7,600,000	3,276,000	3,173,984
Host eggs not parasitized	400,000	356,400	793,496
Parasitized host eggs killed by trauma, percent	80	74	55
Parasitized host eggs killed, number	6,080,000	2,424,240	1,745,691
Parasitized host eggs not killed	1,520,000	851,760	1,428,293
Total parasitized and unparasitized small larvae	1,920,000	1,208,160	2,221,789
Small larvae parasitized, percent	79.2	70.5	64.3
Survival, unparasitized larvae to large larvae, percent	20	22	22
Survival, parasitized larvae to large larvae, percent	15.8	18	18
Unparasitized large larvae	80,000	78,408	174,569
Parasitized large larvae	240,000	153,317	257,092
Total large larvae	320,000	231,725	431,661
Large larvae parasitized, percent	75	66	59.6
Survival large larvae to adults, percent	22.5	23	23
Adult *D. dorsalis*, next generation	18,000	18,034	40,151
Adult *B. arisanus*, next generation	54,000	35,263	59,131
Female *B. arisanus*, next generation[3]	31,482	20,558	34,473

[1]Some arbitrary adjustments are made in reproduction and survival rates due to changes in the host and parasite densities.
[2]See table 1 to calculate the rates of parasitism and escape from parasitism by using the ratio of parasite-host encounters to hosts present.
[3]The ratio of females to males is assumed to be 1.4:1 (58.3 percent females).

the nature described in the basic model shown in table 10. It is assumed that 100,000 female *B. arisanus* will be released per square mile of *D. dorsalis* habitat during the first generation when conditions for reproduction are favorable. The number of parasites in the natural population is ignored. The ratio of adult parasites to adult hosts would be 5:1, or about three times the estimated ratio in a natural population that is assumed to result in 60–65 percent larval parasitism. The values for the various parameters during generation 1 are similar to those in the basic model except that the parasite population is higher. In view of the higher parasite population and the probability of increased predation, the average number of host eggs parasitized by the female *B. arisanus* is assumed to be reduced from 300 for a natural population to 240 for the released population. This assumption is consistent with that made for all the models depicting the influence of augmented parasite populations.

Discussion of Results

The calculations made are largely self-explanatory. For most of the other suppression models developed for this study, the estimated values for the various parameters were based largely on deduction. However, as previously noted, most of the assigned parameter values for the oriental fruit fly model can be related to the extensive ecological data obtained by the authors already cited. Based on the assumed ratio of adult *B. arisanus* to adult *D. dorsalis* and the estimated numbers of eggs deposited by the host females and of parasite-host (-egg) encounters, 95 percent of the host eggs would be parasitized one or more times and 5 percent would escape parasitism. Newell and Haramoto (1968) assumed that when the natural larval parasitization rate averages 75 percent, about 95 percent of the host eggs will have been parasitized and the ratio of adult parasites to adult hosts will be 3:1. If the ratio of females to males averages 1.4:1, as assumed in this model, an approximately 3.5:1 ratio of female parasites to female hosts would cause 95 percent parasitism of the host eggs. However, the model indicates 95 percent parasitism of the *D. dorsalis* eggs when the parasite-to-host ratio is 5:1. Therefore, the estimated efficiency of *B. arisanus* may be very conservative.

Newell and Haramoto observed that the mortality of the parasitized eggs will be about 80 percent when host egg parasitism averages 95 percent. Since the model shows that 95 percent parasitism of the host eggs will result in 75 percent parasitism of the large larvae, an additional mortality factor is required. Introduction of the appropriate factor accounts for the difference between the survival rates of the small parasitized larvae and the small unparasitized host larvae to the large larval stage, the rate being lower for the parasitized larvae.

Based on the model, the *D. dorsalis* population would decline to 9,000 females during generation 2, a decline of 55 percent. However, the *B. arisanus* population would decline by almost 68 percent. Because of these lower densities, slight upward adjustments are made in the average numbers of host eggs produced by the female *D. dorsalis* and of host eggs parasitized by the female parasites. In generation 2, the ratio of adult parasites to adult hosts would decline and so would the rate of egg parasitism. This would result in a decline in the percentage of large parasitized larvae during generation 2. The rate of larval parasitism (66 percent) would not be high enough to cause a further decline in the *D. dorsalis* population during generation 3. Also, the rate of egg parasitism would be reduced even further during generation 3 because of the decline in the ratio of adult parasites to adult hosts. By generation 4 the parasite and host populations would become stabilized, and the rate of parasitism would be near normal. The release of *B. arisanus* for one cycle would cause some decline in the *D. dorsalis* population, but a high degree of control would not occur.

As we shall see, the population and parasitization trends resulting from the release of the egg-larval parasite *B. arisanus* would be very different from those resulting from the release of the larval parasites investigated. While superparasitism of larvae may also cause some early mortality of parasitized hosts, there is no evidence that it is a major mortality factor for most larval parasites. For such parasites, therefore, if the proportion of the parasitized larval population is high enough in the first cycle, the ratio of adult parasites to adult hosts will increase progressively; and this increase will mean a progressive increase in the rate of parasitism. In sharp contrast, the high *B. arisanus* releases will result in a high mortality of the parasitized host eggs; hence, the ratio of parasitized larvae to nonparasitized larvae will be limited. In effect this high mortality is no doubt a natural control mechanism that has evolved to prevent an efficient parasite like *B. arisanus* from completely dominating the host population. If the survival rates of parasite eggs and unparasitized host eggs to the respective adult stages were essentially equal when 95 percent of the host eggs are parasitized, the ratio of adult parasites to adult hosts in the next cycle would be 95:5, or 19:1. But if only 20 percent of the parasitized eggs result in parasitized larvae, the ratio of parasitized larvae to unparasitized larvae would be 19:5. Then, equal survival of the immature parasites and unparasitized hosts would mean a 3.8:1 ratio of adult parasites to adult hosts during the next cycle. In the model it is assumed that a ratio of five adult parasites to one adult host results in 95 percent parasitism of the host eggs. If this is approximately correct, a ratio of

19:1 would certainly result in near-100-percent parasitism. It should be noted again that Newell and Haramoto observed an average of 95 percent parasitism of the host eggs, and some collections indicated 100 percent parasitism when the ratio of adult parasites to adult hosts was about 3:1. Therefore, there is no reason to assume a natural escape mechanism, other than mere chance, that will prevent near-100-percent parasitism of the host eggs if the ratio of parasites to hosts is maintained high enough by artificial means. As discussed in chapter 4, 100 percent of some larval collections of sugarcane borers were parasitized by *Lixophaga diatraeae* after release of this parasite in a sugarcane borer ecosystem. If the ratio of adult parasites to adult hosts is the primary factor that determines the rate of parasitism in natural populations as hypothesized in chapter 2, it would have very great significance in the regulation of pest populations by the parasite augmentation technique.

For parasite species that do not discriminate between unparasitized and already parasitized hosts, the probability of superparasitism depends on the rate of parasitism that occurs. According to Newell and Haramato, the mortality of the parasitized eggs increases as the average rate of egg parasitism increases. This observation is understandable, considering the probability of superparasitism. In accordance with data in table 1, when parasitism by *B. arisanus* averages 95 percent, the host eggs will have been parasitized an average of three times. When parasitism averages 99 percent, the eggs will have been parasitized an average of five times. For delicate host eggs that have been pierced by the ovipositor of a parasite more than once, the probability would seem to be low that many will hatch and result in parasitized larvae. (See also the influence of the egg-larval parasite *Chelonus insularsis* on the parasitism of noctuids, chapter 7).

If excessive mortality of parasitized eggs is the only or primary mechanism for ensuring that the parasite-to-host ratios will not reach levels threatening the survival of the hosts; it could be readily overcome by making supplemental parasite releases in successive life cycles of the host. Such supplementary releases would also be a means of nullifying the protective effect of excessive mortality in superparasitized larvae. In the section to follow, we will analyze the influence of supplemental releases of adult *B. arisanus* on the rates of parasitism of host eggs and larvae.

Supplemental Releases During Several Host Cycles

As the suppression model in table 11 clearly indicates, the high egg mortality limits the ratio of parasites to hosts that can occur. But it would not require many additional parasites to compensate for the mortality that does occur. One can rationalize that hosts in natural populations have never been confronted with highly augmented parasite populations over successive host cycles and so will not have evolved an appropriate protection mechanism. Doubtless, supplementary releases will eventually be technically feasible for most parasite species that suffer high mortality due to a high mortality of the superparasitized hosts.

A separate suppression model is not necessary to indicate the suppressive effects of supplementary releases of *B. arisanus*. Assume that 50,000 additional *B. arisanus* females will be liberated during generation 2 (table 11). The ratio of adult female parasites to adult female hosts would then increase from 5:1 to 9:1. The average number of host eggs parasitized per female may be reduced to some degree because of the lower host density and the rather higher parasite population, but an estimate of 240 eggs per female, as estimated for generation 1, would seem reasonable. In such event, the ratio of parasite-host encounters to host eggs present would be 4.9:1; and this ratio would theoretically result in 99 percent egg parasitism.

If 99 percent of the host eggs were parasitized during generation 2, the host population should decline to a very low level during generation 3. Even so, the ratio of *B. arisanus* adults to the adult fruit fly population would probably be no higher than 4:1. But if 50,000 additional *B. arisanus* were released during generation 3, the ratio of adult parasites to adult hosts would exceed 100:1. I cannot envision any natural defense mechanism that the host population could develop to prevent host-egg parasitism near 100 percent if the parasite-to-host ratio is as high as 100:1. Thus, virtual elimination of the *D. dorsalis* population should occur within a few cycles. The parasite population, in turn, would also be virtually eliminated if *D. dorsalis* were the only available host species. Implementing such a program of initial and supplementary releases would mean that by artificial means, we would be establishing throughout a pest-host ecosystem parasite-to-host ratios that would exceed by manifold the maximum ratios that could occur naturally because of the natural mortality factor. Would it be biologically possible for a host population to survive for long under such circumstances? I think it highly unlikely, but this is the question for which an answer is needed.

Influence of *Biosteres longicaudatus* Releases

Suppression Model

In the development of models to estimate the influence of *B. longicaudatus* (Bl) releases in *D. dorsalis* (Dd) ecosystems, it became apparent that theoretically, such releases would greatly enhance the rate of natural parasitism by *B. arisanus* (Ba). The inherent host-parasitization capability of Bl is estimated to be similar to that of *B. tryoni*, which has already been discussed.

As noted earlier, Bl was one of the more effective parasites for Dd in Hawaii before Ba became established. While its importance as a natural control agent declined because of the dominance of Ba, Bl continues to persist in Hawaii (Wong and Ramadan 1987). Greany et al. (1976) and Lawrence et al. (1978) discussed various aspects of the biology and behavior of Bl. It parasitizes the large host larvae and completes development in the host pupae. The female deposits about 85 eggs. This low fecundity contrasts strongly with that of the female Ba, which deposits more than 700 eggs. As previously noted, however, the fecundity of a parasite species merely reflects the hazards the progeny normally encounter before they reach the adult stage.

Since Ba parasitizes the host eggs and completes development in the host larvae, it apparently preempts successful parasitism by species that parasitize the larval stages. Thus, successful parasitism by Bl in Hawaii will normally be limited to approximately 40 percent of the larvae that have escaped parasitization by Ba. If Bl discriminates and avoids parasitizing larvae already parasitized by Ba, its chances for successful reproduction would be greater than assumed. For this analysis, however, Bl is assumed to not discriminate and to reproduce successfully only in larvae that have escaped parasitism by Ba. If larvae already parasitized by Ba are reparasitized by Bl, the reparasitization is assumed not to influence survival of the Ba larvae. Theoretically, if two parasite species (*a* and *b*) within an ecosystem parasitize different life stages of a coexisting host—with species *a* parasitizing the earlier stages of the host—and if individual hosts parasitized by both species can nevertheless enable the immature stages of species *a* to develop to the adult stage, the augmentative release of species *b* in the ecosystem will result in the enhanced parasitization efficiency of species *a* after the first cycle.

The Bl female is estimated to have the capability of parasitizing 30 host larvae during its lifetime under conditions favorable for reproduction. Thus, in the absence of Ba, if 20 percent of the Bl larvae normally survive to the adult stage, the rate of increase would be 3-fold. However, if 60 percent of the larvae are already parasitized, the rate of increase would be only 1.2-fold. It seems probable that Bl has some discriminating capability and/or is more successful in certain habitats than Ba. Also, its host-parasitization capability must increase substantially when its density is below normal. Otherwise, it most likely could not have survived in competition with Ba.

The model shown in table 12 was developed to indicate the influence of Bl (species *b*) releases in Dd habitats having the coexisting Ba (species *a*) population shown in table 10. The proposal is that 100,000 female Bl be released per square mile during generation 1. The number is the same as the number of female Ba proposed for release in the model depicting the influence of Ba releases.

Discussion of Results

Table 12 is rather involved because it includes a number of parameters that are not needed in estimating the influence of only one parasite species. The values for the parameters relating to Dd and Ba populations during generation 1 are the same as those shown for the basic natural coexisting populations. Since the release of Bl parasites is assumed to have no adverse influence on the development of the Ba population, the Ba population during generation 2 is assumed to be the same as it would be if Bl parasites had not been released. However, the Dd population would be much lower because of the high parasitism by the Bl parasites. Thus, the Ba-to-Dd ratio would increase above the normal ratio.

The release of 100,000 Bl females per square mile against a population of 20,000 Dd females would result in 5:1 ratio of adult parasites to adult hosts. The average number of host larvae parasitized per B1 female is assumed to be 24, a reduction of 20 percent below normal because of increased predation. This arbitrary adjustment is standard for augmented populations. The ratio of total parasite-host encounters to hosts present would be 2.2:1 and would theoretically result in 89.5 percent parasitism. However, since 60 percent of the large larvae would already be parasitized by Ba, only 40 percent would yield potential adult Bl progeny for generation 2. Also, since the potential rate of increase of the Dd population is 2.4-fold, the parasitism caused by Bl would reduce the adult Dd population during generation 2 by about 75 percent. Then, since the Ba population would increase at the normal rate, the ratio of

Table 12
Suppression model depicting the estimated influence of a natural population of *Biosteres arisanus* (Ba) and a released population of *B. longicaudatus* (Bl) in a *Dacus dorsalis* (Dd) ecosystem

Parameter[1] (1 square mile)	Generation		
	1	2	[2]3
Female Dd	20,000	5,103	48
Eggs per female	400	450	600
Total eggs	8,000,000	2,296,350	28,800
Female Ba (natural population)	35,000	84,267	3,805
Ratio, adult Ba to adult Dd	1.75:1	16.5:1	80:1
Eggs parasitized per Ba female	300	185	—
Total Ba-host encounters	10,500,000	15,589,395	—
Ratio, Ba-host encounters to host eggs present[3]	1.3:1	6.8:1	—
Parasitism, percent	73	99	—
Eggs parasitized, number	5,840,000	2,273,387	—
Eggs not parasitized, number	2,160,000	22,964	—
Parasitized eggs killed by trauma, percent	45	95	—
Parasitized eggs killed, number	2,628,000	2,159,718	—
Parasitized eggs not killed, number	3,212,000	113,669	—
Total viable eggs	5,372,000	136,633	—
Survival, viable eggs to large larvae, percent	20	23	—
Large larvae parasitized by Ba	642,400	26,144	—
Large larvae not parasitized by Ba,	432,000	5,282	—
Total large larvae	1,074,400	31,426	—
Large larvae parasitized by Ba, percent	60	83	—
Female Bl released (generation 1)	100,000	43,497	607
Ratio adult Bl to adult Dd	5:1	8.5:1	12.7:1
Large larvae parasitized per Bl female	24	2	—
Total Bl-host encounters	2,400,000	86,994	—
Ratio, Bl-host encounters to large host larvae present[3]	2.2:1	2.8:1	—
Large larvae parasitized by Bl, percent	89.5	92.8	—
Large unparasitized larvae parasitized by Bl	386,640	4,902	—
Large larvae not parasitized by Ba or Bl	45,360	380	—
Survival, large larvae to adults, percent	22.5	25	—
Adult Dd, next generation	10,206	95	—
Adult Ba, next generation	144,540	6,536	—
Adult Bl, next generation	86,994	1,225	—
Adult Ba females, next generation[4]	84,267	3,805	—

[1]Some arbitrary adjustments are made in parameter values due to changing host and parasite densities.
[2]In view of the scarcity of hosts no estimates are made for the rates of parasitism due to Ba and Bl during generation 3.
[3]See table 1 to calculate the rates of parasitism from the ratio of parasite-host encounters to hosts present.
[4]Female-to-male ratio = 1.4:1.

adult female Ba to adult female Dd would theoretically increase from a normal ratio of 1.75:1 to the very high ratio of 16.5:1. Thus, the release of Bl would dramatically change the normal numerical relationship between Ba and Dd.

Based on the theory that kairomones govern the host-finding capability of parasites, as discussed in chapter 2, the assumption is made that the Bl-induced reduction in host population would not cause a comparable reduction in the average host-finding capability of the Ba females. The Ba population, however, has increased by almost 2.5-fold and would be expected to suffer higher mortality as a result of increased predation. Considering both the steep reduction in host density and higher mortality of Ba adults, the average number of host eggs parasitized by Ba females is assumed to be reduced to 185 during generation 2. Also, the average number of host eggs produced by the Dd females is assumed to increase by about 10 percent. But because of the high ratio of adult parasites to adult hosts, the ratio of total parasite-host encounters to host eggs present would be 6.8:1. Theoretically, a ratio of 5:1 results in 99 percent parasitism of the host eggs. However, a maximum of 99 percent parasitism is assumed for all suppression models. A parasitization rate of 73 percent of the host eggs would be expected during generation 2 if Bl had not been released during generation 1. Thus, in theory, the addition of the Bl parasites would not only cause about 90 percent control during generation 1, due to the Bl parasites, but also cause the rate of parasitism of Dd eggs by Ba to increase from about 73 percent to 99 percent during generation 2. The Bl progeny produced during generation 1 should also make a further contribution to controlling Dd during generation 2.

It is difficult to predict the influence of the Ba and Bl populations that theoretically would be present during generation 2, but there is reason to believe that it would be severe. The Dd population would still be fairly high despite a reduction of 75 percent. Such a reduction may have little adverse effect on the average host-parasitization capability of the female Ba present. However, the ratio of adult Ba to adult Dd would theoretically increase from a normal of 1.75:1 to 16.5:1. For a normal population as shown in table 10, the average number of eggs parasitized per Ba female during generation 2 is assumed to be as high as 270. For this analysis, it is assumed that the average number parasitized will be only 185 eggs. Even so, in view of the high parasite-to-host ratio, 99 percent of the host eggs would theoretically be parasitized by Ba during generation 2. The total population of large larvae would therefore decline from a normal of 1,074,000 to 31,426. Of this number, only 5,282 are estimated to escape parasitization by Ba. The adult Bl female population would theoretically exist at a level of 43,497. So if the Bl females parasitized as few as two larvae each, the rate of parasitism of the unparasitized larvae would be 92.8 percent.

Based on probable conservative parameter values during generation 2, the Dd population would decline to 48 females during generation 3. Then, in theory, the ratio of adult Ba to adult Dd during generation 3 would be 80:1, and that of adult Bl to adult Dd would be 12.7:1. What influence such grossly distorted ratios would have on the rates of parasitism when the host population exists near the level of extinction is the question for which an answer is needed. Such an answer is needed not only for tropical fruit flies but for all other insect pests that may be candidates for suppression by the parasite augmentation technique.

Of special importance are the results supporting the earlier mentioned theory that the efficiency of a parasite species can be enhanced by the augmentative release of another species that parasitizes a later life stage of a host common to the two species. If such an enhancement of parasitism is confirmed as a biological phenomenon, it might expand the potential of the parasite augmentation technique. It would mean that substantial advantage could be gained by integrating the release of two or more parasite species. Theoretically, the principles of suppression that become involved would go beyond parasite releases. Suppression of a pest population by any target-pest-specific technique (which may involve using, for example, sterile insets, sex pheromones, or pest-specific chemicals) could lead to enhanced biological control by natural populations of host-specific parasites, because the normal ratio of adult parasites to adult hosts would increase. Such enhanced parasitism could be expected if, as hypothesized, the average number of hosts parasitized by the adults in a parasite population does not decline proportionally to the decline in the number of available hosts. These theoretical possibilities could hardly have been envisioned without the use of suppression models of the type developed for this investigation.

In conclusion, table 12 clearly indicates that the release of Bl, or a similar species that parasitizes the larval stages, would be more effective than the release of Ba. The high mortality of the superparasitized host eggs will limit the ratio of adult Ba to adult hosts that can coexist in the generation following the generation that the releases are made. While superparasitism of host larvae may also reduce the number of parasite progeny produced, the reduction would seem to be relatively minor for Bl or any other species that parasitize fruit fly larvae. In any event, as previously discussed, any such reduction could be nullified by supplemental releases of parasites in a series of host cycles.

Although this discussion was focused on Dd, the same principles and biological actions discussed should apply to the medfly, because this fruit fly is also parasitized by Ba and Bl.

Potential of *Biosteres vandenboschi* as a Species for Release

As noted earlier, *B. vandenboschi* (Bv) proved to be an effective parasite of *D. dorsalis* (Dd) and the medfly before competition by *B. arisanus* (Ba) reduced its effectiveness. Parasitism data obtained by Wong and Ramadan (1987) indicate that Bv is now scarce in Hawaii.

According to Vanden Bosch and Haramoto (1953), Bv prefers to parasitize first-instar larvae. Its fecundity of about 160 eggs suggests that its inherent host-parasitization capability is about one-fifth that of Ba and about two times that of Bl. These comparative fecundity data suggest that the normal survival rate of Dd host eggs to the larval stage parasitized by Bv is about 15 percent and that the normal survival rate of the small larvae to the large larval stage is about 50 percent. These are the types of clues relied upon to estimate values for key parameters used in population models.

Simulation models will not be presented, but I estimate that under favorable conditions, the Bv female will parasitize an average of about 65 host larvae during its lifetime. If 10 percent of the parasite larvae normally survive to the adult stage, the rate of increase of Bv would be about 3.25-fold. However, if 60 percent of the small larvae are already parasitized by Ba and if Bv cannot discriminate or compete with Ba, its reproduction would be greatly reduced. Doubtless, an inability to discriminate or to compete accounts for the great decline in Bv populations after Ba became the dominant fruit fly parasite in Hawaii.

Despite the disadvantage of Bv in ecosystems containing natural populations of Ba, simulation models show that the release of Bv at the rate of about 100,000 females per square mile would have essentially the same influence as that described for the release of Bl. Theoretically, Bv releases would greatly enhance the rate of parasitism by the natural Ba population during generation 2 for the same reasons described for Bl releases.

Thus, if Bv could be mass-produced more readily and inexpensively than Bl, it may be a more practical species for suppressing Dd populations than Bl. However, each species may have advantages on certain Dd host fruits or in certain habitats. Therefore, the release of both species may be more effective than the release of either species alone.

Scenario of a Program To Eradicate the Oriental Fruit Fly in Hawaii and an Estimate of the Costs

While suppression models may indicate the potential effect of a parasite species when released in a pest-host ecosystem, the results in operational programs cannot be expected to equal the theoretical results. A number of variables exist that cannot be readily evaluated in suppression models. Therefore, to achieve satisfactory results, the number of parasites required may have to be severalfold greater than the theoretically adequate number. Also, the models are based on discrete generations; but for tropical fruit fly species, all of the life stages will be present when parasite releases begin.

For the scenario of a Dd eradication program in Hawaii, the Dd population is assumed to average 20,000 adults of each sex per square mile, and the area harboring Dd is assumed to total 3,000 square miles. Thus, I estimate that 100,000 female Bl (or Bv) would have to be released per square mile of Dd habitat during each of three successive generations. Parasite releases would be started when two criteria are met: the host population is low and conditions for reproduction would soon become favorable. Theoretical suppression models indicate that releases in one generation only would lead to near elimination of a Dd population within three generational cycles. Nevertheless, for this program, sterile fly releases are assumed to be necessary for Dd eradication.

The adult Dd population by generation 4 is assumed to average 1,000 males and 1,000 females per square mile. Beginning with that generation, the sterile flies would be released during each of three successive generations at an average rate of 100,000 of each sex per square mile. In generation 4, this release rate would result in an overall sterile-to-fertile ratio of 100:1, but an effective ratio of 25:1 is assumed. Theoretically, this ratio would reduce the natural population to an average of 400 native flies of each sex in the next generation even if the increase rate of those reproducing were 10-fold. This reduction is based on the assumption that the parasites present would contribute no further control. The sterile fly releases during generation 5 would then increase the effective sterile-to-fertile ratio to about 62:1. By generation 6, the sterile-fly releases should result in such high sterile-to-fertile ratios that the probability of a fertile mating in a 1-square-mile area would be zero.

According to this scenario, a total of 600,000 parasites and 600,000 sterile Dd would have to be released per square mile of Dd habitat. The costs of rearing and releasing *B. tryoni* were estimated to eventually total $2,000 per million. This is also assumed to be the cost of rearing and releasing Bl or Bv adults. The cost of rearing and releasing sterile Dd adults is estimated at $600 per million. Therefore, the cost per square mile for the insects required for the hypothetical program described would be $1,560. For 3,000 square miles of Dd habitat, the cost would be $4,680,000. Assuming that the costs for surveys prior to and during operation and the overall costs of supervision and administration would aggregate $2,320,000, the total cost for the program would be about $7 million. This would be only about one-tenth the preliminary estimates of the probable cost for eradication that is based largely on the use of insecticides (U.S. Department of Agriculture 1984).

The estimate of $7 million required for the hypothetical eradication program may be overly high. Theoretical suppression models indicate that parasite releases for only one generation would lead to near elimination. The estimate is based on parasite releases for three generations. Also, only 95 percent reduction in the Dd population is assumed by generation 4. The possibility that enough parasite progeny will be present to cause additional control during generations 4–6 is also disregarded. Thus, several very conservative assumptions are made.

On the other hand, the natural Dd population may well exceed the estimated average of 20,000 males and 20,000 females per square mile during the generation of lowest density. Also, natural balancing factors may come into play with sufficient intensity or unknown natural escape from parasitism may occur to prevent the degree of parasitism that the suppression models indicate.

It should be kept in mind, however, that the eradication scenario and cost estimates are based on the release of parasites alone for three cycles, followed by the release of sterile flies. As previously theorized, the suppressive actions of sterile flies and parasites would be mutually potentiated if these biological agents were released concurrently. This theory of mutual potentiation was first presented by Knipling (1979), and it is also discussed in greater detail in chapter 3. The concurrent release of sterile flies and parasites beginning with generation 1 would be much more cost effective than the kind of program described. Other possibilities exist for minimizing the requirements and costs. The total area where Dd reproduction occurs permanently may be considerably smaller than the areas of temporary spread each year. In theory, complete inhibition of reproduction only in the areas of permanent survival for 1 year would lead to elimination from temporary survival areas, where full advantage could be taken of natural control factors.

Thus, although a number of negative factors might exist that would require more insects than predicted by the theoretical models, a number of positive factors might permit successful eradication at even lower costs than estimated. Several new principles of pest population suppression not previously recognized might become involved when releases of different parasite species and sterile flies are appropriately integrated.

In general, this theoretical analysis indicates a very high probability that a major fruit fly species, Dd, can be eliminated from Hawaii by highly practical and ecologically acceptable procedures. There would seem to be ample justification for undertaking the very difficult and demanding research and pilot experiments that would be required to fully evaluate the different techniques and strategies suggested by this theoretical analysis.

Suppression of Caribbean Fruit Fly Populations

The Caribbean fruit fly (Carib fly), *Anastrepha suspensa* Loew, is a pest of subtropical fruits in Florida. While citrus is not a primary host, it is sufficiently vulnerable to *A. suspensa* infestations to create a quarantine problem. Citrus and certain other fruits intended for shipment to sensitive areas must be treated and certified for export. The fruit industry in Florida is interested in eradicating or suppressing the pest so that fruits would be allowed to move through the marketing channels without treatment.

The parasite *Biosteres longicaudatus* (Bl), which has already been discussed as a candidate for suppressing oriental fruit fly populations, is effective against the Caribbean fruit fly in Florida. According to R.M. Baranowski (University of Florida, Homestead, personal communication), natural parasitism of Carib fly larvae ranges up to 50 percent or more. Various aspects of the biology and behavior of Bl have been investigated by Greany et al. (1976), Lawrence et al. (1976), Greany et al. (1977), and Lawrence et al. (1978).

Using previously described procedures for estimating the potential of the parasite augmentation technique to suppress the oriental and Mediterranean fruit flies in Hawaii, I analyzed the potential of Bl releases to suppress populations of *A. suspensa* in Florida. The pest distributes itself throughout much of Florida during the warmer months but retreats to southern parts of the State during winter because of cold weather and absence of suitable hosts (R.M. Baranowski, personal communication). This retreat offers the opportunity for concentrating eradication or suppressive measures in areas of permanent survival of the pest and relying on natural control factors for eradication in temporary survival areas. In the sections to follow, the possibility of suppressing *A. suspensa* populations by the augmentation technique and achieving eradication by the sterility technique will be considered. The late Loren Steiner (U.S. Department of Agriculture, unpublished report) demonstrated that low-density populations of *A. suspensa* can be eliminated by the release of sterile flies. The field experiments were conducted in Florida during the early 1970's. The Florida State Experiment Station and the U.S. Department of Agriculture have investigations under way on sterile fly and/or parasite releases for suppressing the Carib fly. Whether the use of sterile flies alone will be economically feasible as an eradication measure is not known. The use of insecticides to suppress populations to manageable levels by the sterility technique would be unacceptable, especially in residential areas, because of public concern over potential hazards of insecticide sprays. Threfore, suppression of *A. suspensa* to low levels by the augmentation technique could represent an important contribution to an eventual effective and ecologically acceptable solution to the *A. suspensa* problem in Florida.

Natural Coexistence Pattern of Caribbean Fruit Fly and *Biosteres longicaudatus* Populations in Florida

The model shown in table 13 is assumed to be representative of the actual and relative numbers of *A. suspensa* and Bl existing in natural populations in Florida during the winter and spring generations. The key parameters are the relative numbers of hosts and parasites in typical populations and the relative numbers of large larvae produced by the female hosts and of large larvae parasitized by the coexisting female parasites. The values for these parameters and the survival rates to the respective adult stages determine the population and parasitism trends. The *A. suspensa* population is assumed to average 10,000 males and 10,000 females per square mile during generation 1.

Table 13
Basic model depicting the estimated coexistence pattern of the Caribbean (Carib) fruit fly and *Biosteres longicaudatus* (Bl) populations in Florida during the late winter and spring

Parameter[1] (1 square mile)	Generation		
	1	2	[2]3
Female Carib flies	10,000	30,000	54,000
Large larvae per female	50	40	32
Total large larvae	500,000	1,200,000	1,728,000
Female Bl	6,666	20,000	36,000
Ratio, adult parasites to adult hosts	1:1.5	1:1.5	1:1.5
Host larvae parasitized per female	30	24	20
Total host larvae parasitized	200,000	480,000	720,000
Percent parasitism	40	40	42
Host larvae not parasitized	300,000	720,000	1,008,000
Survival, unparasitized host larvae and immature parasites to adults, percent	20	15	10
Total Carib flies, next cycle	60,000	108,000	100,800
Total parasites, next cycle	40,000	72,000	72,000
Increase rate of host population	3.0	1.8	1.0
Increase rate of parasite population	3.0	1.8	1.0

[1]Arbitrary adjustments are made in certain parameter values because of increasing host and parasite densities.
[2]Populations are assumed to decline to normal low levels during periods unfavorable for reproduction.

It is estimated that under favorable conditions, the female *A. suspensa* produces an average of 50 large host larvae during its lifetime. The female Bl is assumed to have the capability of parasitizing an average of 30 host larvae during its lifetime. This value is the same as the estimated average number of oriental fruit fly larvae Bl parasitizes under favorable conditions in Hawaii (see section on oriental fruit fly suppression). The averages are assumed to decline as the parasite and host populations increase to higher densities. Natural larval parasitism due to Bl in Florida is assumed to average near 40 percent. If the survival rates of the immature parasites and unparasitized host larvae to the respective adult stages are similar, as assumed, the ratio of adult parasites to adult hosts would typically be near 1:1.5.

Doubtless, Bl is responsible for preventing *A. suspensa* populations from reaching much higher levels. However, like other tropical fruit fly parasites, it is not capable of controlling the host satisfactorily for human needs, because of nature's balancing mechanism. Thus, for adequate biological control, the assumption is that adults of one or more key parasite species must be increased in the Carib fly ecosystems by artificial means.

Influence of *Biosteres longicaudatus* Releases

The basic model shown in table 13 was used to estimate the influence of Bl releases throughout the parts of the pest-host ecosystem that permit year-round survival of the pest. The suppression model is shown in table 14. The parameter values for the host population during generation 1 are similar to those assigned for a natural population. Enough Bl are assumed to be released to achieve a 4:1 ratio of adult parasites to adult hosts. Based on an estimated natural population of 10,000 *A. suspensa* of each sex per square mile, the release required would be 80,000 parasites of both sexes per square mile. The natural population of Bl is ignored.

Because of the abnormally high parasite population, predation will likely increase and result in a shorter average life of the parasites.

Table 14
Suppression model depicting the estimated influence of *Biosteres longicaudatus* (Bl) releases against a Carib fly population of the nature shown in table 13

Parameter[1] (1 square mile)	Generation		
	1	2	3
Female Carib flies	10,000	7,250	2,778
Large host larvae per female	50	50	55
Total host larvae	500,000	362,500	152,790
Female Bl released	40,000	42,750	35,284
Ratio, adult parasites to adult hosts	4:1	5.9:1	12.7:1
Host larvae parasitized per female	24	22	[2]18
Total parasite-host encounters	960,000	940,500	635,112
Ratio, parasite-host encounters to host larvae present[3]	1.9:1	2.6:1	4.1:1
Parasitism, percent	85.5	92.7	98.3
Host larvae parasitized, number	427,500	336,038	150,193
Host larvae not parasitized	72,500	26,463	2,597
Survival rates to adults, percent	20	21	23
Total Carib flies, next generation	14,500	5,557	597
Total Bl, next generation	85,500	70,568	34,544

[1]Some arbitrary adjustments in parameter values are made because of the declining host population.
[2]The larval host population would exist about 70 percent below the normal low level. According to the influence of kairomone guidance mechanisms, as discussed in chapter 2, the average host-parasitization capability of the female parasites would be reduced a maximum of about 25 percent.
[3]See table 1 to calculate the rates of parasitism and escape from parasitism by using the ratio of parasite-host encounters to hosts present.

Thus, the average host-finding capability of the females is assumed to be reduced from 30 to 24. The results of the releases are largely self-explanatory. The rate of parasitism during generation 1 would theoretically reach a level of about 85.5 percent. If the survival rates of the immature parasites and unparasitized host larvae to the adult stage are approximately equal, as assumed, the ratio of adult parasite progeny to adult host progeny would be about 5.9:1 during generation 2. The rather small reduction in the host population during generation 1 is not likely to have an appreciable effect on the average host-finding capability of the parasites during generation 2. However, because of some decline in the host population and a slightly higher parasite population, the average number of hosts parasitized is assumed to be reduced by about 10 percent, from 24 to 22 per female. The magnitude of this reduction would be much smaller than the magnitude of increase in parasite-to-host ratio. Therefore, the ratio of parasite-host encounters to hosts present would increase substantially and result in a higher rate of parasitism. Theoretical parasitism would be 92.7 percent. This, in turn, would lead to a greater-than-double increase in the parasite-to-host ratio during generation 3. The larval host population would theoretically decrease by about 70 percent. I estimate that the average number of host larvae parasitized per female would be reduced by 25 percent. If so, the ratio of parasite-host encounters to hosts present would be 4.1:1; and this ratio would theoretically result in 98.3 percent parasitism.

Like all theoretical suppression models, the accuracy of the values assigned to the various parameters may be open to question. The most important is the estimated average host-parasitization capability of the female parasites because of the declining host density. However, as discussed in chapter 2, there is overwhelming evidence that the reduction in host-finding capability of the female parasites will not be proportional to the decline in host density below the normal low level. Even by generation 3, the larval host population would probably be within the range of host densities that are normal under unfavorable conditions for reproduction. It is questionable whether the average host-parasitization capability would be reduced significantly when the larval density is about 70 percent below the estimated normal low level. But the assumption is made that the average host-parasitization capability will be 25 percent below normal. Nevertheless, because of the high ratio of adult parasites to hosts, the rate of parasitism would theoretically increase to 98.3 percent, and the ratio of adult parasites to adult hosts would then increase to 58:1 during generation 4.

A factor not considered in the calculations is the survivability of developing immature parasites in hosts that are heavily reparasitized. Lawrence et al. (1978) found that superparasitism will result in a lower survival rate of the immature parasites. However, this finding is based on a laboratory study in which female parasites had the opportunity to reparasitize the host larvae many times. At the parasitism level of 85.5 percent shown in the model, the larvae would be parasitized an average of about two times. During generation 2, when parasitism is estimated to be 92.7 percent, the average would be about 2.6 times. In generation 3, the average would be 4.1 times. I doubt if the survival rate would be greatly reduced except, possibly, by generation 3, when the rate of parasitism is estimated to be 98.3 percent. If the survival rate of the immature parasites were reduced to half the normal rate during generation 3, the ratio of adult parasites to adult hosts during cycle 4 would still be high, 29:1, and the rate of parasitism would increase. If the survival rate of parasites in superparasitized hosts is lower than estimated, supplemental releases of relatively few adult parasites can be made in each cycle.

Potential Advantage of Concurrent Releases of Parasites and Sterile Flies

The sterile fly technique provides control by limiting the successful reproduction of native females; the parasite augmentation technique, by limiting the survival of the host larvae. From the standpoint of basic principles of insect population suppression, the concurrent use of the two techniques would be more advantageous than the use of either technique by itself, as discussed in some detail in chapter 3.

Table 15 presents the theoretical results of an integrated attack on a Carib fly population. The Carib fly population is assumed to exist as depicted in table 13. Only enough sterile flies, namely 80,000 flies of both sexes, are released during generation 1 to achieve an overall sterile-to-fertile ratio of 4:1. If the flies are only 25 percent competitive, the effective ratio would be 1:1. Therefore, only 50 percent of the native females would reproduce successfully. The release of 12,500 female Bl, or less than one-third the number proposed for parasite releases alone, would result in a 2.5:1 ratio of adult parasites to adult reproducing hosts. The release of about 25,000 parasites of both sexes per square mile is assumed to result in some increase in predation. This inrease and the reduction in host density may reduce the average host-finding capability of the female parasites, so the female parasites are estimated to parasitize an average of 26 host larvae. Theoretical parasitism would be 73

Table 15
Suppression model depicting the theoretical effect of concurrent releases of sterile Carib flies and *Biosteres longicaudatus* (Bl) parasites in a Carib fly ecosystem having a population of the nature shown in table 13

Parameter (1 square mile)	Generation		
	1	2	3
Female Carib flies	10,000	7,088	726
Sterile males released[1]	40,000	40,000	40,000
Effective sterile to fertile ratio	1:1	1.4:1	13.8:1
Control due to sterile flies, percent	50	58	93.2
Female Carib flies reproducing	5,000	2,977	49
Host larvae produced per female[2]	50	53	60
Total host larvae	250,000	157,779	2,940
Female Bl released, generation 1 only	12,500	19,162	17,056
Ratio, adult Bl to reproducing Carib flies	2.5:1	6.4:1	348:1
Host larvae parasitized per female[3]	26	23	1
Total parasite-host encounters	325,000	440,726	17,056
Ratio, parasite-host encounters to host larvae present[4]	1.3:1	2.8:1	5.8:1
Parasitism, percent	73	94	99
Host larvae parasitized, number	182,500	148,312	2,911
Host larvae not parasitized	67,500	6,311	29
Survival to adults, percent[2]	21	23	33
Adult Carib flies, next generation	14,175	1,452	9
Adult Bl, next generation	38,325	34,112	961

[1]The sterile flies are assumed to be only 25 percent competitive.
[2]Arbitrary adjustments are made in parameter values because of a declining host density.
[3]Because of the high parasite population and the lower host density, the average number of hosts parasitized per female parasite is assumed to be below normal. If each female parasitizes an average of only 1 host larva during generation 3, the theoretical rate of parasitism would still exceed 99 percent.
[4]See table 1 to calculate the rates of parasitism and escape from parasitism by using the ratio of parasite-host encounters to hosts present.

percent. Fifty percent control due to sterile flies and 73 percent control due to the parasites would aggregate 86.5 percent control during generation 1. This rate of control would be higher than that (85.5 percent) estimated to result from the release of more than three times the number of parasites alone. About eight times as many sterile flies alone would be required to cause 88 percent parasitism.

The influence of the integrated attack would increase during generation 2. The supplementary release of 80,000 sterile flies would increase the effective sterile-to-fertile ratio relatively little. However, the ratio of parasites to reproducing female hosts would increase to about 6.4:1. The average number of host larvae parasitized by the female parasites is assumed to decline but not enough to compensate for the higher parasite-to-host ratio. Theoretical parasitization is estimated to increase to 94 percent. The combined effect would be about 97.5 percent control during generation 2.

The full impact of the insect releases would occur during generation 3. Another supplementary release of 80,000 sterile flies would result in an overall sterile-to-fertile ratio of about 55:1, or an assumed effective ratio of about 13.8:1. The ratio of adult parasites to reproducing host females would theoretically increase to 348:1.

We are, of course, confronted with the usual question of how host-parasitization capability is affected by a host population that is near extinction. But assuming that the average capability would be only one larva per female, theoretical parasitism during generation 3 would be above 99 percent. The Carib fly population by generation 4 would be near elimination. The ratio of adult parasites to adult hosts would be about 100:1. We cannot predict what contribution the parasites would make to suppression when the host density is virtually extinct; however, we can predict, without question, the influence of the sterile fly releases. Theoretically, the release of 80,000 sterile flies per square mile during generation 4 would result in an overall sterile-to-fertile ratio of more than 8,000:1.

The theoretical data in the model show that the suppressive actions of the two techniques would be mutually enhanced, in accordance with the discussion in chapter 3. Whether such enhancement can be achieved has not yet been tested in experimental or operational programs. But such an enhancement would seem to be especially desirable in tropical fruit fly eradication programs. Obviously, theoretical models can only indicate the principles involved and the extent to which the two techniques may be complementary. Field trials will be required to confirm the basic principles that become involved.

Scenario of a Program To Eradicate the Caribbean Fruit Fly in Florida and an Estimate of the Costs

As previously noted, suppression models of the nature developed for this theoretical study have little practical significance unless they can be related to the probable requirements, costs, and benefits of operational programs. One of the most uncertain assumptions is the number of Carib flies in Florida and how they are distributed. Nevertheless, I believe that gross estimates can be used to indicate the magnitude of the requirements and costs likely to be involved.

In attacking the Carib fly population in Florida by the techniques proposed, it is my view that they need to be applied to only those Carib fly habitats that permit year-round fly reproduction. If reproduction can be inhibited in these areas for a full year, natural control factors should prevent survival in areas where reproduction occurs only during favorable periods of the year. For the purpose of this analysis, the assumptions are that permanent Carib fly habitats total 5,000 square miles and that the parasite releases will begin in the late winter or early spring, when the host population is near the minimum level but the Bl population is able to survive and reproduce.

The theoretical model in table 14 indicates that parasite releases resulting in a 4:1 ratio of adult parasites to adult hosts during the first generation would cause the rates of parasitism to increase such that by the end of generation 3, the Carib fly population would be nearly eliminated without supplemental control. It may be presumptuous, however, to assume that the releases would be as effective in an operational program. A more realistic estimate of probable requirements is that the releases of Bl be made for three successive host cycles. Based on an initial average host population of 10,000 of each sex per square mile, 80,000 parasites of both sexes would have to be reared and released per square mile for three cycles, or a total of 240,000 parasites per square mile. If this release rate is necessary on 5,000 square miles of Carib fly habitats, the total requirements for Bl would be 1.2 billion.

Fortunately, the technology for rearing Carib flies and the parasites is already well advanced because of research conducted jointly by the University of Florida and the U.S. Department of Agriculture. We might assume that Bl can eventually be mass-produced in well-designed rearing facilities and released in large numbers at a cost of $2,000 per million. On this basis, the cost of rearing and releasing 1.2 billion parasites would be $2.4 million. Table 14 indicates that by generation 4, the Carib fly population would be reduced to less than 1,000 adults per square mile if the parasites are released for only one generation. The model based on concurrent releases of parasites and sterile flies (table 15) involves fewer parasites and a relatively low ratio of sterile flies to fertile flies during the first and second generations. However, theoretical elimination would occur during generation 4.

For the purpose of this appraisal, the number of sterile flies required is assumed to equal the number of parasites, namely, 80,000 per square mile per generation. The cost of rearing and releasing the sterile flies is estimated to be $600 per million, or a total cost of $0.72 million for 1.2 billion sterile flies. Thus, the cost of the insect components would aggregate about $3.1 million (2.4 + 0.72 = 3.12). Additional costs would be involved, however, to allow for adequate surveys and surveillance and for supervision, administration, and contingencies. Hence, the total costs would likely increase to about $6 million, not including the cost of facilities.

The estimates may be too low in certain aspects, but they may be too high in others. Appropriate pilot experiments would have to be undertaken to test the technology and the most effective strategies. This theoretical appraisal indicates, however, that in Florida, the parasite release technique appropriately integrated with sterile fly releases would be cost effective over the long term in eliminating isolated Carib fly populations and should greatly benefit the Florida fruit industry. Since both techniques are target-pest specific, environmental hazards would be avoided. This is a major consideration in coping with tropical fruit fly problems, especially in residential areas.

Can Parasite Releases Eradicate Infestations of Newly Introduced Tropical Fruit Flies?

This subject of eradicating infestations of newly introduced tropical fruit flies is posed as a question because no information is yet available to permit a positive or negative answer. However, one of the major concerns of fruit and vegetable industries in tropical and subtropical agricultural areas is the accidental introduction and establishment of a number of species that are capable of causing hundreds of millions of dollars' worth of losses and that may require for their control the use of broad-spectrum insecticides, which have the potential for causing serious environmental damage. Consequently, drastic and often costly measures are undertaken to eliminate infestations when discovered. Based on the theoretical analysis of the role of natural parasite populations as biological control agents and the role they might play when released in tropical fruit fly ecosystems, the question raised deserves serious consideration by scientists and pest management officials having responsibilities relating to tropical fruit fly problems.

The possibility that parasite releases could be employed for eradicating newly established populations of these pests will be analyzed from a broad perspective as well as in considerable depth. The analysis will involve a specific, hypothetical case; and consideration will be given both to the various factors that influence the insect parasitization process in natural environments, as discussed in chapters 2 and 3, and to the theoretical results of parasite releases, as discussed earlier in this chapter.

As evident from their very existence for many thousands of years, the primary parasites of specific hosts or host complexes have the inherent capability of finding and parasitizing enough hosts to maintain viable populations even when the host populations drop to abnormally low levels. As noted elsewhere, the ability of a parasite species to readily find its hosts when the host population is low and widely dispersed is probably as vital to its survival as the ability of insects to find mates and continue reproducing under such circumstances. Also, as theorized, the role such parasites play as natural control agents depends primarily on the actual and relative numbers of parasites and hosts that can coexist in natural populations because of the natural balancing forces. Regarding new infestations of tropical fruit flies, we can make reasonably accurate estimates of their actual numbers. It should be technically feasible to distort the normal numerical ratio in favor of the parasite population by a factor of up to 1,000 or more when host populations are very low. These are the broad, basic premises of the analysis.

The medfly will be used in the analysis as the hypothetical, newly discovered host, and its actual numbers and distribution patterns will be estimated. But any of a wide range of tropical fruit fly species or other kinds of insect pests could be used for the analysis.

When a new medfly infestation is discovered, it is likely to be in about the third cycle, even in an area with a good surveillance program. It is also likely to have originated from a single reproducing female. If conditions for reproduction are favorable, the potential rate of increase is likely to be as high as 10-fold per generation. Therefore, one reproducing female would produce about 10 male and 10 female progeny. These numbers would, in turn, increase to about 100 males and 100 females, and then to about 1,000 males and 1,000 females, and so forth. So early discovery of new infestations is important, regardless of the eradication methods that might be employed.

In this hypothetical case, we will assume discovery of the infestation when 1,000 male and 1,000 female medflies will be emerging and then reproducing during a period equivalent to one cycle. Most of the reproduction is likely to be in an area of about 1 square mile or less, but scattered reproduction can also be expected to occur in an area of 25 square miles. Each female at such a low density is assumed to produce an average of 80 large host larvae during its lifetime. Then, a 25-percent survival rate of the larvae to adults would result in 20 adult progeny per female and account for the 10-fold increase rate. Of the total larval host population of 80,000, isolated populations of about 80 larvae (each population originating from 1 female) could be expected to be widely distributed in the total 25-square-mile area, and a few populations of about 800 larvae could be expected to exist in some isolated habitats. Thus, we have assigned absolute values for the key parameters that relate to the medfly population.

Biosteres tryoni is assumed to be the parasite released, although several other species may also be suitable. The release of two or more species would likely ensure greater success than the release of only one species. In the hypothetical case, an important assumption made is that no other acceptable host for *B. tryoni* exists in the infested area. *B. tryoni* has a strong affinity for medfly larvae and normally parasitizes the large larvae. As discussed earlier, normal favorable conditions will enable the female parasite to parasitize an estimated average of 30 host larvae during its lifetime. If the survival rate of the immature *B. tryoni* is comparable to the estimated 25-percent survival rate of the unparasitized host larvae, the female parasites would each produce an average of 7.5 adult progeny and, thus, increase the population 3.75-fold. In view of the very low host density and the distribution pattern, a lower-than-average host-parasitization capability for the female parasites is to be expected.

The assumption is made that 1 million male and 1 million female *B. tryoni* will be released during a period comparable to 1 medfly generation. Releases will be made somewhat in accordance with the available information on the density and distribution of the infestation. However, because the exact locations of most of the populations constituting the infestation will not be known, a large number of the parasites will be distributed at random throughout the 25-square-mile area. The release of 1 million females would mean an average of 40,000 females per square mile, or 62 per acre. The ratio of adult *B. tryoni* to adult medflies would be an overwhelming 1,000:1 overall but would vary according to host distribution within the 25 square miles.

While most of the *B. tryoni* would be released in host-free habitats, the parasites are assumed to actively search for host habitats, taking advantage of kairomone signals emanating from infested fruit, as described by Greany et al. (1977) for a related species. Nevertheless, in view of the widely scattered low-density host habitats, only 10 percent, or 100,000, of the released females are assumed to be successful in finding host habitats. Also, the number of host larvae parasitized by the successful female is assumed to average 10, rather than 30, the number assumed for more normal host densities. On the basis of these assumptions, 100,000 females could parasitize 1 million larvae. However, the larval population totals only 80,000. Therefore, the ratio of the total number of parasite-host encounters to the total number of hosts available would be 12.5:1. Theoretically, a ratio of 5:1 will result in 99 percent parasitism, or the escape of only 1 percent by chance. A maximum of 99 percent parasitism is assumed for all models, however.

In general, I feel that plausible assumptions and estimates have been made in analyzing the potential of the parasite release technique under the hypothetical conditions described. A parasitization rate of 99 percent when the increase rate is 10-fold would reduce the population to 100 males and 100 females during the next medfly cycle. The release of another 1 million female *B. tryoni* would then lead to a 10,000:1 ratio of adult parasites to adult hosts. The same rate of parasitism would then reduce the population to 10 males and 10 females; thus, a subsequent release of 1 million female *B. tryoni* would result in an overall ratio of 100,000:1.

The scenario described indicates a good possibility that the release of parasites can be a means of eradicating infestations of newly introduced medflies. It is recognized, however, that assumptions and estimates are made that cannot be related to experimental data or field observations. The assumption that the rate of parasitism can reach a 99-percent level may be questioned. Yet, as discussed earlier, Newell and Haramoto (1968) reported that 95 percent of oriental fruit fly eggs in natural populations were parasitized when the ratio of adult parasites to adult hosts was estimated to be 3:1. In our hypothetical case, we are dealing with overall ratios as high as 1,000:1 or higher.

The overall parasite-to-host ratios resulting from parasite releases are impressive, but they may not be significant unless related to the actual numbers and spatial distribution of medfly larvae in new areas of spread. It is probable that in dozens of isolated locations, only one female medfly will be reproducing and that, as noted in chapter 3, the larval progeny will be concentrated in an area of less than 1 acre. The search area during the life of a *B. tryoni* female might be assumed to be about 5 acres. At the parasite release rate proposed, approximately 300 females could be expected to be present in 5 acres. If so, the ratio of adult parasites to adult hosts would be 300:1 in areas where a single female host is reproducing. When viewed from this perspective, the numerical relationship created by the release would be very impressive. In a natural population, a 1:1 ratio of adult parasites to adult hosts can be expected to result in about 50 percent parasitism. What the rate of parasitism would be when the ratio is 100:1, 1,000:1, or even higher in an area where a single host is reproducing is the question for which an answer is needed.

The sterile insect technique has proven to be effective in eliminating newly introduced low-level medfly infestations and in preventing the establishment of populations in areas subject to the infiltration of flies from nearby established populations. However, the parasite release method may have several advantages over the sterility technique, as discussed in some detail in chapter 3. The released parasites should be more normal in vigor and behavior because irradiation damage would not be involved. Also, they can be expected to search actively for host habitats. In contrast, sterile male and female insects released together into the environment cannot be expected to search selectively for nonsterile mates.

While the costs involved in eliminating newly discovered tropical fruit fly infestations are usually not a major concern, they are still an important consideration. The hypothetical case discussed is based on the release of 6 million parasites of both sexes during a period corresponding to three medfly cycles. At the previously estimated cost to eventually mass-produce and release *B. tryoni* ($2,000 per million), the cost of the parasites needed in the hypothetical case would be $12,000. In contrast, a single insecticide spray application on 25 square miles would likely cost $25,000. If a minimum of six applications were required, the total would be $150,000. If even as many as 30 million parasites prove to be needed for the successful eradication of a medfly infestation like the one discussed, the parasite release method could make a major contribution to tropical fruit fly eradication technology. Of particular advantage would be the use of an eradication method that would cause no health or environmental problems.

As difficult and challenging as it might be to conduct an investigation to determine whether parasite releases can eliminate new isolated fruit fly infestations, this theoretical analysis indicates that there would be justification for giving such an investigation very high priority.

7 Suppression of Fall Armyworm Populations in the Florida Overwintering Areas

The fall armyworm (FAW) *Spodoptera frugiperda* (J.E. Smith) is representative of insect pests that have low populations and a greatly restricted range of distribution during the winter months. The populations grow in spring and summer, when conditions favor reproduction, and may spread for hundreds of miles during the crop-growing season. Such pests are capable of causing extensive damage before cold weather again pushes them back to their normal overwintering areas. This is the pattern for FAW, as noted by Luginbill (1928), Vickery (1929), and Snow and Copeland (1969). South Florida and south Texas are the areas of permanent survival in the United States.

A logical strategy for controlling important pests having such behavior is to apply suppressive pressure against the total population during the winter so that the number of insect pests emigrating to non-overwintering areas would be too small to be economically damaging. The possibility of managing the FAW population in the South and Southeast according to this strategy was discussed by Knipling (1980). Several techniques were suggested that might accomplish the objective of suppressing the overwintering populations, including the release of sterile or partially sterile moths, the use of sex pheromones to inhibit reproduction, the release of biological agents, and the application of appropriate cultural measures.

During the past decade, excellent progress was made by research scientists on several techniques of suppression that have their greatest efficiency when the target pest populations are low. Carpenter et al. (1983) showed that when FAW moths are subjected to low dosages of irradiation, a high percentage of their progeny will be sterile. Sex pheromones of FAW, which have been described by Tumlinson et al. (1984), would be useful for survey purposes and may be effective for suppression (Mitchell and McLaughlin 1982). In this chapter, the potential for suppressing populations of FAW by the parasite augmentation technique will be appraised.

FAW is attacked by several parasite species that would seem to be good candidates for augmentation purposes. A promising species is the egg-larval parasite *Chelonus insularius* (Cresson) (fig. 5).

Vickery (1929) recorded an average of 33 percent parasitism of FAW larvae in south Texas due to this hymenopterous species. Parasitism reached a maximum level of about 65 percent. Parasitism of FAW by *C. insularius* (Ci) is also prevalent in Florida. Ashley et al. (1983) recorded average parasitism near 40 percent, with a high level of about 61 percent. Information is lacking on actual numbers of the parasites and hosts that are present per unit area and the rates of parasitism. Therefore, the model to be described is based largely on estimates.

Natural Coexistence Pattern of Fall Armyworm and *Chelonus insularius* Populations in Florida

The hypothetical model shown in table 16 is proposed as representative of the actual and relative numbers of the FAW and Ci that coexist during late winter through early spring in Florida and the rate of parasitism that prevails.

While FAW has a wide range of host plants, A.N. Sparks (retired), E.R. Mitchell, and J.R. Young of the U.S. Department of Agriculture's Agricultural Research Service believe (personal communication) that sweet corn, field corn, and volunteer corn are the host plants primarily responsible for FAW reproduction during the winter and early spring in Florida. The total acreage of cultivated and wild host plants during the winter probably does not normally exceed 50,000. Estimates of 50 million moths during the period corresponding to generation 1 and 100–150 million during generation 2 are used for planning purposes. These estimates would mean an average of 500 females per acre of the assumed acreage of host plants during generation 1 and more than 1,000 during generation 2. These numbers of moths could readily account for extensive damage to host plants if no control measures were applied. In my opinion, the total population estimates are likely to be representative of "worst-case" situations in the Florida overwintering areas.

Knipling (1980) estimated that 5 million immigrant moths from the overwintering area during the spring would produce nearly 4 billion moths in the non-overwintering areas by season's end. These estimates were based on an assumed average growth rate of 5-fold per life cycle for four cycles. If 250 females per acre during a period equal to 1 generation causes significant damage, the estimated moth populations would mean that about 8 million acres of crops would be at risk. This could account for the high losses the pest causes in some years. A substantial proportion of the moths

Figure 5. *Chelonus insularius.* This insect is one of the most effective of the natural biological agents for the fall armyworm, an important pest of grain and forage crops. *C. insularius* parasitizes host eggs, and the immature parasites complete their development in the host larvae. This parasite causes rates of larval parasitism up to 50 percent or higher in some cases. Even so, it does not adequately control the fall armyworm. An appraisal of the effectiveness of *C. insularius* suggests that if natural populations of the parasite could be increased severalfold by artificial means, the rate of parasitism would be high enough for adequate control. The fall armyworm is representative of a number of insect pests that have a limited distribution and relatively low populations during the winter months but that spread throughout much of the United States in spring and summer. The release of reared parasites in the restricted overwintering areas is proposed as a means of greatly reducing the number of emigrating moths. Such a reduction would prevent populations in the non-overwintering areas from increasing enough to cause significant damage before cold weather again pushes the pest back to its overwintering areas. (Photograph, courtesy of W. Joe Lewis, USDA Insect Biology and Management Systems Research, Tifton, GA.)

Table 16
Basic model depicting the estimated coexistence pattern of the fall armyworm (FAW) and the egg-larval parasite *Chelonus insularius* (Ci) in the overwintering areas in Florida

Parameter (1 acre of host plants)	Generation 1	Generation 2
FAW females	500	1,200
Eggs per female[1]	333	300
Total eggs	166,667	360,000
Ci females	300	780
Ratio, adult parasites to adult hosts	.6:1	.65:1
Host eggs parasitized per female	400	360
Total parasite-host (-egg) encounters	120,000	280,800
Ratio, parasite-host encounters to host eggs present[2]	.72:1	.78:1
Host eggs parasitized and superparasitized, percent	52	55
Host eggs parasitized, number	86,667	198,000
Host eggs not parasitized, number	80,000	162,000
Parasitized eggs killed, percent	40	42
Parasitized eggs not killed	52,000	114,840
Total number of eggs that can hatch	132,000	276,840
Percent of larvae parasitized	39.4	41.5
Survival rates to adult stage, percent[1]	3	2.5
Adult FAW, next generation	2,400	4,050
Adult Ci, next generation	1,560	2,871
Increase rate of FAW population	2.4	1.7
Increase rate of Ci population	2.6	1.8

[1]Arbitrary changes in values are made during generation 2 because of the higher densities.
[2]To calculate the rate of parasitism from the ratio of parasite-host encounters to hosts present, see table 1.

developing each cycle during the warmer months must disperse up to 100 miles or more. There is no other explanation for the damage the pest often causes before season's end in areas up to 500 miles or more from the overwintering areas.

The size of the host population can be used to estimate the size of the coexisting Ci population. Based on parasitism data obtained by Ashley et al. (1983), a larval parasitization rate of 40 percent during the winter and spring would be representative of normal parasitism. A basic assumption in this investigation is that for solitary larval parasites, the rates at which immature parasites and unparasitized host larvae survive to the adult stage are approximately equal. Therefore, if larval parasitism averages 40 percent during a given cycle, the ratio of adult parasites to adult hosts during the next cycle should be near 40:60. Thus, if the adult FAW population averages 500 females per acre, the adult Ci population should average about 333 females per acre. For the coexistence model I assume 300 per acre during generation 1.

Other key parameters are the average number of host eggs produced by the FAW females during their lifetime and the average number parasitized by the coexisting Ci females. A number of estimates for these key parameters were considered before accepting the relative values shown. Vickery (1929) determined that the FAW female is capable of depositing an average of about 1,000 eggs. W.J. Lewis (U.S. Department of Agriculture) kindly provided unpublished data on the fecundity of Ci females. The average is about 1,200 eggs per female. As noted elsewhere (chapter 2), the average number of eggs deposited by insects in natural populations is estimated to be about 33 percent of the fecundity under normal favorable conditions for survival. This rule-of-thumb estimate was helpful in developing the model. The actual numbers of eggs produced and eggs parasitized may deviate considerably from the estimates of 400 for Ci and 333 for the FAW females, but the numerical relationship between these parameters should be fairly accurate. For most parasite-host complexes, the fecundity of the host exceeds that of the associated parasite. If the reverse is true, one can assume that the parasite faces more formidable life hazards than does the host.

The model was designed to simulate a host-population growth rate of about 2-fold during the first generation and a larval parasitization rate of about 40 percent. The estimated ratio of adult parasites to adult hosts in the model is 0.6:1. Parasitism data for Ci are based on the estimated rate of larval parasitism, but the species parasitizes the host eggs and completes development in the larvae. If the estimated parasite-to-host ratio and the relative average numbers of host eggs produced and of host eggs parasitized are approximately correct, the rate of host-egg parasitization must be considerably higher than the observed rate of larval parasitization. For the model to reflect 40 percent larval parasitization, a high mortality factor is required for the parasitized eggs. No data are available to indicate that such mortality occurs. However, as discussed earlier, Newell and Haramoto (1968) observed that the egg-larval parasite *B. arisanus* causes considerable mortality of oriental fruit fly eggs in the parasitization process. They also observed that the higher the rate of egg parasitization, the higher the mortality of the parasitized eggs. In the process of developing the model, I reached the conclusion that Ci also causes a high mortality of host eggs in the parasitization process and that there is a positive correlation between the mortality and the percentage of the host eggs parasitized.

Based on the average numbers of host eggs produced and host eggs parasitized by the coexisting FAW and Ci females, the ratio of total parasite-host encounters to eggs present would be 0.72:1. Theoretically, then, 48 percent of the eggs will escape parasitization and 52 percent will be parasitized at least once. Since the rate of larval parasitism is assumed to average near 40 percent, the mortality rate of the eggs parasitized must be on the order of 40 percent. A number of different values were considered for the key parameters in developing the model. The values shown are considered the most realistic. The FAW population increases by 2.4-fold during generation 1. The Ci population increases by 2.6-fold. Slight adjustments are made in the average numbers of eggs produced and eggs parasitized during generation 2, because of the higher host and parasite densities. Also, the rates of survival of the immature parasites and large unparasitized host larvae to the respective adult stage are assumed to decrease for the same reason. Thus, the rates of increase of the populations during generation 2 would be somewhat lower than those during generation 1. Ashley et al. (1983) observed an apparent increase in parasitism as the density of the FAW population increased. But such an increase does not necessarily mean that host density is positively correlated with the rate of parasitism. An increase in parasitism could mean that a higher proportion of the FAW population than the Ci population emigrated from the habitats surveyed. The parasite-to-host ratio in the overwintering areas would then increase and, as a result, the rate of parasitism would also increase. The model shows a rather slight increase in the rate of parasitism during generation 2.

Parameter values in models of the nature shown may deviate somewhat from true values. As examples, the average numbers of eggs produced and eggs parasitized may be higher or lower than assumed, and the survival rates may also be correspondingly lower or higher than indicated in the model. However, the relative values cannot deviate much without causing unrealistic relative trends in the populations and unrealistic rates of parasitism within as few as two or three cycles. The estimated mortality of the parasitized eggs is a key parameter in the model. As will be considered in greater detail in the next section, a high mortality factor for the parasitized eggs must occur; otherwise, an efficient parasite like Ci would, in time, completely dominate the host population. Without such a mortality, it is questionable whether the two organisms could coexist. Indeed, a high mortality of parasitized and superparasitized eggs may be characteristic of all species of egg-larval parasites and perhaps some larval parasites. If so, this natural control factor could be readily overcome by making supplemental parasite releases. In the next section, we will consider the theoretical influence of parasite releases for several successive generations.

Influence of *Chelonus insularius* Releases

Suppression Model

The basic model in table 16 was used with some slight modifications to calculate the influence of Ci releases for one generation throughout the FAW overwintering areas, as depicted in table 17. The primary objective of such a program of releases would be to reduce the emigrating population to so low a level that economically damaging populations would not develop in the usual areas of spread before the end of the crop-growing season. Protection of crops in the overwintering areas would be a secondary objective.

As indicated, a typical population of FAW in Florida during the period designated as generation 1 is assumed to consist of 50 million moths of both sexes. The coexisting population of Ci is assumed to number 30 million. The proposal is to release enough Ci throughout the host ecosystem during generation 1 so that the parasite population is raised to 150 million. The parasite-to-host ratio would then increase to 3:1—or approximately five times the estimated normal ratio, which results in about 40 percent larval parasitism. As shown in table 16, a parasite-to-host ratio of 0.6:1 is estimated to cause 52 percent parasitism of the host eggs and 39.4 percent parasitism of the host larvae. A 5-fold increase in the parasite population achieved by artificial means throughout the FAW ecosystem could be expected to greatly increase the rate of parasitism and reduce the dynamics of the FAW population. The presence of the abnormally high parasite population is assumed not to affect the average egg-laying potential of the FAW females. However, the high Ci population is assumed to result in higher predation and a somewhat lower average number of host eggs parasitized by female Ci. The ratio of adult parasites to adult hosts is so high, however, that the ratio of the total number of parasite-host encounters to the total eggs available would be 2.9:1. This will theoretically result in 94.6 percent parasitism of the host eggs. Only the eggs that escape parasitism have the potential to produce FAW progeny. Thus, allowing for the expected normal increase rate, the FAW population would be reduced by about 68 percent.

A highly significant influence of the parasite releases would be the high mortality of the parasitized eggs. Such a mortality would reduce the ratio of parasitized to nonparasitized host larvae and,

Table 17
Suppression model depicting the estimated influence of *Chelonus insularius* (Ci) releases in fall armyworm (FAW) ecosystems (see text for more details)

Parameter[1] (1 acre)	Generation		
	1	2	3
FAW females	500	158	84
Eggs per female	333	366	400
Total eggs	166,667	57,828	33,600
Ci females (released generation 1)	1,500	469	203
Ratio, adult parasites to adult hosts	3:1	3:1	2.4:1
Host eggs parasitized per female	320	320	300
Total parasite-host encounters	480,000	154,070	60,900
Ratio, parasite-host encounters to host eggs present[2]	2.9:1	2.6:1	1.8:1
Percent host eggs parasitized	94.6	92.7	84
Host eggs parasitized, number	157,667	53,607	28,224
Host eggs not parasitized, number	9,000	4,221	5,376
Host eggs and young larvae killed by parasitism, percent[3]	83	81	70
Parasitized eggs and young host larvae killed, number	130,864	43,422	19,557
Parasitized eggs and young host larvae not killed, number	26,803	10,185	8,467
Total parasitized and unparasitized larvae	35,803	14,406	13,843
Host larvae parasitized, percent	75	70.7	61.2
Survival rates to adults, percent	3.5	4	4.5
Adult FAW, next generation	315	169	242
Adult Ci, next generation	938	407	381

[1]Arbitrary adjustments are made in estimated values for the average egg deposition, average number of eggs parasitized, and the survival rates, because of the declining host and parasite densities.
[2]See table 1 to convert ratio to rate of parasitism.
[3]The estimates are based, in part, on data obtained by Newell and Haramoto (1968) for a similar egg-larval parasite.

hence, limit the ratio of adult parasites to adult hosts in the next generation. As noted, the lower FAW population during generation 2 is assumed to result in some increase in the average number of eggs deposited by the female hosts and a lower average rate of egg parasitism by the female parasites. But these advantages would be too slight to nullify the continuing high parasite-to-host ratio. The result would be a further reduction in the FAW population during generation 2, but the ratio of parasites to hosts and the rate of egg parasitism would begin to decline.

Discussion of Results

According to the model, the FAW population will decline by about 83 percent by generation 3. Thus the population would be about 96 percent lower than an unsuppressed population in generation 3. Such a reduction in population should cause a proportionate reduction in the number of emigrating moths. Therefore, the release of the parasites for one generation only may achieve the objective. But the FAW population would recover near normal reproductive success within about four generations because of the egg mortality factor. By generation 3, the number of adult progeny would begin to exceed the parent population, and near normal larval parasitism could be expected by about generation 5. These results contrast strongly with those theorized for the release of parasites that parasitize the host larvae directly. For most larvae-seeking parasites, moderate superparasitism seems to cause little or no mortality of the host larvae before the parasites complete development. Therefore, a high initial parasite-to-host ratio can theoretically lead to progressively higher parasite-to-host ratios and progressively higher rates of parasitism for several successive generations. This phenomenon has already been discussed in other sections of this report.

If the unparasitized host eggs and the parasite eggs deposited in host eggs survived to the adult stage at the same rate and if, as the model indicates, the rate of parasitism of the host eggs during generation 1 were on the order of 94.6 percent, the ratio of adult parasites to adult hosts in the next cycle would be about 94.6:5.4, or 17.5:1. A reduction of about 68 percent in the FAW egg population would probably cause little or no reduction in the average host-parasitization capability of the Ci females. In such event, the total number of parasite-host encounters would exceed the total number of host eggs present by 10-fold or more. This would mean that less than 1 percent of the host eggs would escape parasitism by chance and also that 99 percent of the eggs would be parasitized one or more times. Then, if the survival rates of the parasite eggs and unparasitized host eggs continued to be equal, the ratio of adult parasites to adult hosts during generation 3 would be 99:1. Thus, the significance of the high mortality of the parasitized eggs becomes apparent. It limits the rate of larval parasitism and the ratio of adult parasites to adult hosts that coexist in the population in subsequent cycles. This is regarded as a unique natural mechanism to avoid possible elimination of a host population by a highly efficient parasite.

Table 17 shows that 83 percent of the parasitized eggs will be killed when the rate of egg parasitism is 94.6 percent. Thus, the ratio of parasitized to nonparasitized larvae would be about 16.1:5.4, or approximately 3:1. This would mean 75 percent larval parasitism. Then, if the survival rates of the immature parasites and unparasitized larvae to the respective adult stages are equal, as assumed, the ratio of adult parasites to adult hosts during generation 2 would again be 3:1 instead of 17.5:1. The FAW population has declined considerably, however, and so has the Ci population. The average number of host eggs deposited by the female FAW is assumed to increase by about 10 percent because of the lower density. The average number of eggs parasitized by the Ci female is assumed to remain at 320. Thus, the ratio of parasite-host encounters to eggs present would decrease, and the decrease would result in slight declines in the rates of parasitism and of parasitized-egg mortality. Therefore, the ratio of surviving parasitized larvae to the unparasitized host larvae would decline. The percentage of larvae parasitized during generation 2 is estimated to be 70.7 percent, a substantial decline from 75 percent during generation 1. The parasitization rate in generation 2 would mean a considerable decline in the ratio of adult parasites to adult hosts during generation 3, and a further decline in the rate of parasitism of the host eggs would be expected.

The key factor in the model is the estimated rates of mortality of the parasitized eggs. The estimates are based largely on the findings of Newell and Haramoto (1968) on the egg-larval parasite *Biosteres arisanus*. They estimated that when 95 percent of the oriental fruit fly eggs are parasitized, the parasitized-egg mortality will be about 80 percent and the rate of larval parasitism will be about 75 percent. However, by deduction I estimate that the actual mortality will be about 84 percent. They observed that the mortality of the eggs increases as the rate of parasitization increases. This is understandable. When parasitism averages 95 percent, the hosts will be parasitized an average of about three times. When parasitism reaches 99 percent, the hosts will be parasitized an average of about five times. A logical assumption is that the mortality of the delicate eggs will increase with an increase in the average number of times the eggs are parasitized.

The high mortality of the parasitized eggs is an excellent example of behavioral effects that are detrimental to individuals in a population but beneficial for the welfare of the population as a whole. The increased mortality with increased egg parasitism ensures that the ratio of adult parasites to adult hosts will not reach a level that will threaten survival of the host. Since high egg mortality has evolved as a natural regulating mechanism, it tends to support the hypothesis on which much of this investigation is based, namely, that the ratio of adult parasites to adult hosts largely determines the rate of parasitism. Concerning Ci and *B. arisanus*, I doubt that larval parasitism can greatly exceed 75 percent regardless of the rate at which the host eggs are parasitized. Thus, the ratio of adult parasites to adult hosts is never likely to exceed about 3:1 in natural populations.

Such a natural regulating mechanism may also occur to some degree among some of the species that parasitize larvae. However, for others, superparasitism does not necessarily lead to an increase in the mortality of the superparasitized larvae or the immature parasites. Indeed, in some species, more than one parasite progeny may occasionally be produced by superparasitized larvae. Also, for species that can discriminate—a common behavior for some species (Van Lentern 1981)—the probability of excessive superparasitism would be minimized. Thus, I believe that among most parasite-host associations, the danger of excessive natural parasitism is avoided in other ways. As discussed in chapter 2, natural balancing forces tend to maintain ratios of parasites to hosts in natural populations at a constant level or within narrow limits. It is my opinion that for most larval parasites, there is no danger that natural parasitism will approach or exceed an average of 50 percent. In fact, a cursory review of recorded rates of parasitism for many parasite-host complexes indicates that average parasitism for most species is well below 50 percent. The two egg-larval parasite species examined in this study, particularly *B. arisanus*, seem to be exceptions. But as noted, they have evolved a mechanism that will ensure that the ratio of adult parasites to adult hosts will not progressively increase beyond a certain level. Fortunately, the science of insect pest management has the capability of maintaining, through artificial means, parasite-to-host ratios at levels that will exceed ratios in natural populations by manifold. In the following section, we will consider the theoretical influence of supplemental Ci releases that will more than compensate for the egg mortality that results from superparasitism.

Influence of *Chelonus insularius* Releases in Successive Host Generations

If a high mortality of superparasitized eggs is a natural mechanism for limiting the parasite-to-host ratio that can develop in natural populations, as the theoretical analysis clearly indicates, the species that have evolved this mechanism would be highly vulnerable to suppression by releases of large numbers of the adult parasites in successive host generations. We will consider this vulnerability by referring to generations 2 and 3 in the suppression model (table 17). What defense would the FAW population have if on the order of 1,000 female Ci were released during generation 2 to supplement the 469 females that develop naturally? The average host-finding capability of the females present should not be reduced by much, if any, despite the reduction in host density. The addition of the parasites throughout the host ecosystem would result in a higher density and probably a higher degree of predation. But adequate allowance can be made for this negative factor. For this analysis, we might assume that 1,500 Ci females per acre would coexist with 158 female FAW during generation 2. Hence, the parasite-to-host ratio would be about 9.5:1, as compared with a ratio of 3:1 resulting from naturally produced progeny. If the females are assumed to be capable of parasitizing an average of at least 200 host eggs during generation 2, the ratio of total parasite-host encounters to total eggs present would be at least 300,000:57,828, or 5.2:1. Theoretically, this ratio would result in 99 percent parasitism of the eggs and probably on the order of 97 percent mortality of the parasitized eggs. Such a mortality would mean that the ratio of adult parasites to adult hosts would still remain near 3:1 during generation 3. However, because very few eggs would escape parasitism, the FAW population would be suppressed to about 16 females per acre during generation 3. Then, the release of as few as 500 additional female Ci per acre would result in a parasite-to-host ratio exceeding 31:1. In theory, the local FAW population may be virtually eliminated. But enough moths traveling long distances from other sources would no doubt immigrate into and repopulate the area. Thus, elimination would not be biologically possible. However, by having released Ci for two generations, the early-season FAW population and, hence, the emigrating population would theoretically be reduced by 99 percent or more.

The possibility of egg parasitism as high as 99 percent may be questioned. However, as noted elsewhere, Newell and Haramoto (1968) observed an average of 95 percent natural parasitism of oriental fruit fly eggs and nearly 100 percent parasitism in some collections when the ratio of adult *B. arisanus* to adult hosts was estimated to be 3:1. Since we would theoretically be achieving Ci-to-adult-FAW ratios of about 9:1 during generation 2 and a much higher ratio during generation 3 if additional parasites were released, the probability of 99 percent parasitism of the eggs should not be discounted.

Cost Analysis

If the release of Ci during generations 1 and 2 reduced the emigrating FAW population by no more than 95 percent, it would seem unlikely that FAW populations throughout the usual areas of spread could increase to the extent that they would cause significant damage before the end of the crop-growing season. Thus, the objective would have been achieved. If we accept the estimate of 50 million FAW moths overwintering in Florida, the cost of the Ci parasites required can be estimated within a reasonable range of error. Because the suppression model is based on a parasite-to-host ratio of 3:1 during generation 1, the release of 150 million Ci adults would be necessary. To overcome the natural mortality factor, 100 million additional Ci are proposed for release during generation 2. The cost of rearing and releasing a total of 250 million Ci may not be very high. Mass-rearing of the parasite was undertaken many years ago by Tardrew (1951) and Bedford (1956). With the technological advances that have been made in mass-rearing insects, it would seem conservative to assume that the cost of rearing and releasing Ci would not likely exceed $5 per 1,000, or $5,000 per million. Thus, the cost of rearing and releasing 250 million Ci should not exceed $1.25 million. Obviously, such a program would have to be supported and guided by thorough surveys. Also, in operational programs, at least two times the theoretically adequate number of insects would no doubt have to be reared and released to compensate for unavoidable inefficiencies. Therefore, the cost of rearing and releasing the insects might be assumed to amount to $2.5 million. Even if the cost of surveys, supervision, and administration increased the estimated cost to $4 million, the program would still be considered highly successful if it reduced the economic losses due to the pest to a level of no significance. While the losses due to FAW seem to vary greatly from year to year, the average losses over a period of years would doubtless exceed by 10-fold or more the cost of such a preventive program. Additionally, the reduction of economic loss would be accomplished by an ecologically acceptable control procedure. The only alternative at present is to apply ecologically disruptive insecticides when crops are subject to damage.

While, in theory, the release of parasites would alone be highly suppressive, a major contribution to pest management might be made at relatively low cost by the application of appropriate cultural measures. The destruction of sweet corn stalks immediately after the ears are harvested and the control of volunteer corn in all fields may alone greatly reduce the size of the overwintering FAW population. Also, the release of relatively few FAW moths treated for optimum inherited sterility effects could contribute significantly to the suppression program. Thus, there is good reason to believe that technology can be developed to resolve the FAW problem in the Southeast at a fraction of the cost of losses that the pest now causes under current control practices. However, until the insect-pest-management community and the agricultural segments that would benefit are willing to develop and employ appropriate technology for preventive programs, growers will have to continue relying on reactive control measures. These measures currently involve the application of ecologically disruptive chemical insecticides. The theoretical analysis presented indicates that Ci augmentation offers outstanding potential as a practical and acceptable way of suppressing FAW populations in the overwintering areas in Florida and hence of protecting crops from this pest each year throughout the eastern seaboard region.

Parasite augmentation may have a great advantage over other suppressive measures. As discussed elsewhere, the movement of highly mobile parasites may make their release in small unisolated areas impractical. However, when such parasites are released in adequate numbers throughout the pest-host ecosystem, their ability to readily move into the host habitats could be one of the greatest assets of this method of control. We can be certain that Ci has the capability of readily locating its host habitats, even though noctuids generally are also highly mobile. Without this capability, Ci would not be the successful parasite that it is.

8 Suppression of Gypsy Moth Populations

Overview

A major effort has been made for years to introduce parasites and other biological agents that would control the gypsy moth, *Lymantria dispar* L. (U.S. Department of Agriculture 1981). Several such agents have become established and have, no doubt, alleviated the damage the pest causes. However, the introduced species developing on their own have been incapable of preventing periodic outbreaks of the moth in long-established areas or in preventing the moth from spreading to new areas. Consequently, the gypsy moth causes continuing losses to the forest industry. It is also an important pest in forested recreational areas and is of major concern to owners of shade trees in residential areas.

Natural balancing mechanisms maintain relative numbers of parasites and hosts and predators and prey within certain limits. The nature of the pest and the plants or animals affected determine whether the balancing mechanisms will favor the biological agents or the pest. Many potential insect pests are effectively controlled by natural biological agents. But others are not, and if they are to be controlled biologically, it is my view that the biological agents must be increased by artificial means. I consider the gypsy moth to be in the latter category of pests.

Several species of parasites might be considered promising candidates for controlling the gypsy moth by augmentation releases. The most important requirement in selecting a well-adapted species for augmentation purposes is likely to be the availability of technology for rearing large numbers of the species at reasonable cost.

In an earlier publication, the potential of *Brachymeria intermedia* to control the gypsy moth was evaluated by theoretical means (Knipling 1977). It is my view that this hymenopterous pupal parasite may be very effective for augmentation purposes, despite discouraging results by Blumenthal et al. (1979). The parasites were liberated in small unisolated areas and failed to achieve significant parasitism. However, as previously stated, the release of a highly mobile species in unisolated areas consisting of a few hundred or even a few thousand acres is not likely to result in adequate parasitism. We know relatively little about the dispersal

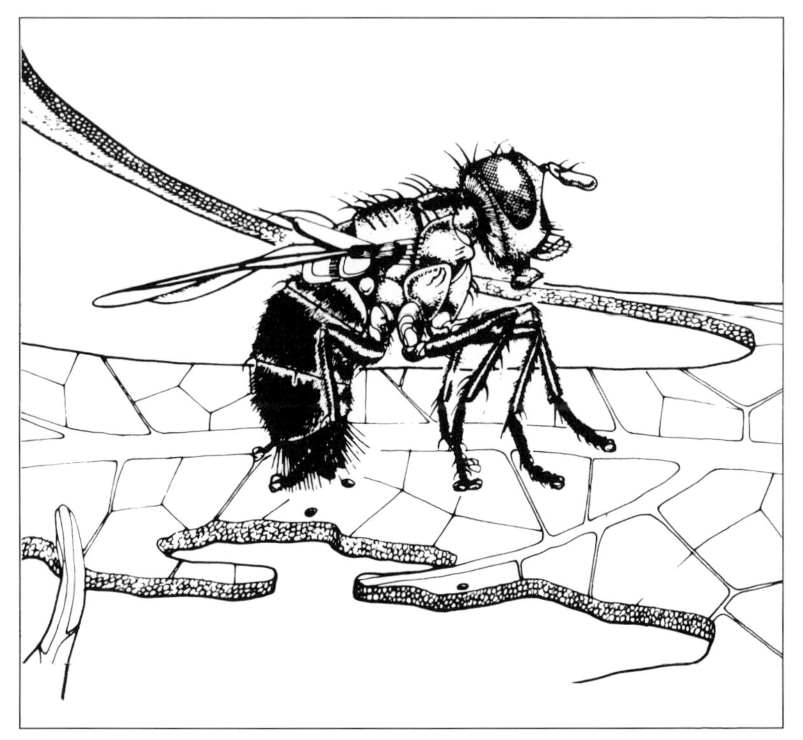

Figure 6. *Blepharipa pratensis.* This tachinid parasitizes larvae of the gypsy moth, an alien species that seriously damages timber and shade trees. The immature parasites complete their development in gypsy moth pupae. *B. pratensis* is one of the most prevalent of the parasites that have been introduced to control the gypsy moth. Like natural populations of many other parasites, however, those of *B. pratensis* cannot increase to levels that will prevent periodic outbreaks of the gypsy moth or prevent its spread to new areas. Theoretical calculations suggest that releasing adequate numbers of *B. pratensis* adults when gypsy moth populations are still low would prevent outbreaks that would cause damaging defoliation. Controlling the gypsy moth by a gypsy moth-specific parasite would allay public concern over possible environmental hazards associated with the use of nonselective insecticides. (Drawing, courtesy of Thomas M. Odell, USDA Forest Service.)

behavior of most parasites. We do know, however, that some species are capable of dispersing for many miles. The host density may influence the dispersal behavior. Grimble (1975) conducted an experiment to study the movement of *B. intermedia* that he released in a small area having a rather high gypsy moth population. The results he obtained contrast strongly with those obtained by Blumenthal et al. Grimble collected gypsy moth pupae at various distances from the release area. Parasitism on the order of 32 percent was recorded. The rate of parasitism about 800 meters from the release area was about equal to the rate of parasitism near the release area. Some parasitism was recorded well beyond 800 meters.

The results obtained by Grimble were similar to those obtained by Summers et al. (1976) for a tachinid, *Lixophaga diatraeae*, that parasitizes larvae of the sugarcane borer (see discussion on the sugarcane borer, chapter 4). Most of the dispersion of *L. diatraeae*, which produces several generations per season, was due to the progeny of the released parasites. The data obtained by Summers et al. indicate that the efficiency of parasitism was higher than the theoretical estimate made by Knipling (1972). On the other hand, when the number of parasites released by Grimble is related to the size of the area of spread and the rate of parasitism achieved, the efficiency determined experimentally and the efficiency I estimated by theoretical procedures seem to generally agree.

The purpose of this analysis is to estimate the potential of the tachinid parasite *Blepharipa pratensis* for use in augmentation programs (fig. 6). This larval-pupal parasite seems to be one of the most prevalent of the gypsy moth parasites in Europe and in the United States (U.S. Department of Agriculture 1981). Considerable information has been obtained on its biology and behavior (Godwin and Odell 1981). It seems to be specific for the gypsy moth. The females deposit their eggs on the foliage of host trees. The eggs must be ingested by the larger gypsy moth larvae to cause parasitism. *B. pratensis* has an unusually high fecundity, depositing up to 4,820 eggs (Godwin and Odell 1981). The high fecundity is indicative of the hazards the parasite population faces before adult progeny are produced.

As will be considered later, *B. pratensis* is an interesting subject for theoretically estimating and then comparing the relative rates of parasitism that could be expected (1) when the parasite eggs are deposited randomly on leaves and (2) when the eggs are deposited only on leaves within habitats harboring host larvae. Such an analysis can help us appraise the importance of host-guidance mechanisms on the host-finding capability of parasites.

Natural Coexistence Pattern of Gypsy Moth and *Blepharipa pratensis* Populations

Coexistence Model

The model shown in table 18 is proposed to represent the relative trends of naturally coexisting populations of *B. pratensis* and the gypsy moth and the rate of parasitism that occurs as the gypsy moth population increases from a low, stable endemic level to an outbreak level. Outbreaks of the host population are assumed to occur sporadically. The gypsy moth is probably more variable in its distribution in the environment than are most other insect pests. The number of egg masses per unit area will likely vary greatly from habitat to habitat when populations exist at stable endemic levels. However, for this analysis, the egg-mass density is assumed to average 20 per acre in a large gypsy moth ecosystem. When conditions for reproduction become highly favorable, the gypsy moth population increases and spreads to adjacent areas, causing considerable defoliation, before it again subsides as a result of the total effect of various density-dependent suppression forces. Natural parasitism due to *B. pratensis* is assumed to average 20 percent. This assumed average parasitism is consistent with certain data reported by Semevsky (1973) and by the U.S. Department of Agriculture (1981).

The primary purpose for developing the model was to estimate the actual and relative numbers of parasites and hosts that are likely to coexist when the host population is in an outbreak mode. Appropriate values must be estimated or assumed for several parameters in the simulation model, including the average number of large host larvae produced by the gypsy moth females and the average number parasitized by the coexisting female parasites during their lifetime. The relative survival rates of the immature parasites and unparasitized host larvae to the respective adult stages during successive generations will then indicate the actual and relative rates of growth of the two populations and the parasitism trend. As discussed in chapter 2, the rates at which the immature parasites within the parasitized hosts and the unparasitized host larvae survive to the respective adult stages are assumed to be similar for many parasite-host complexes. The actual survival rates can be expected to decline as the host and parasite densities increase, but the relative survival rates are assumed to remain rather constant

Table 18
Basic model depicting the estimated coexistence pattern of gypsy moth and *Blepharipa pratensis* populations when the gypsy moth population is in an outbreak mode

Parameter (1 acre)	Generation (year)				
	[1]1	2	[2]3	[2]4	[2]5
Egg masses	20	160	957	3,636	5,256
Large host larvae (4th–6th instar) per mass[3]	160	136	109	82	41
Total host larvae	3,200	21,760	104,313	298,152	215,496
Female *B. pratensis*	5	40	240	1,058	2,203
Ratio, adult female parasites to egg masses	1:4	1:4	1:4	1:3.4	1:2.4
Host larvae parasitized per female	128	109	98	83	50
Total host larvae parasitized	640	4,360	23,520	88,131	110,150
Percent parasitism	20	20	22.5	29.6	51
Host larvae not parasitized	2,560	17,400	80,793	210,021	105,346
Survival to adult stage, percent[3]	12.5	11	9	5	2
Adult gypsy moths, next generation	320	1,914	7,271	10,511	2,107
Adult *B. pratensis*, next generation	80	480	2,117	4,407	2,203
Egg masses, next generation[4]	160	957	3,636	5,256	[5]1,054
Increase rate of host population	8	6	3.8	1.4	.2
Increase rate of parasite population	8	6	4.0	2.1	.5

[1]The gypsy moth population is assumed to be in an outbreak mode, and the average number of large larvae per egg mass and their survival rate during generation 1 are similar to those given in the life table reported by Campbell (1981).
[2]The increase in the rate of parasitism starting in generation 3 is due to a greater percentage decrease in average number of host larvae produced per egg mass than in the average number of host larvae parasitized per female parasite, and not due to an increase in the number of hosts parasitized because of the increasing host density.
[3]As the host and parasite populations increase, some adjustments are arbitrarily made in the average numbers of host larvae produced and host larvae parasitized and in survival rates.
[4]Female moths generally outnumber males. However, some mortality of females can be expected before eggs are deposited, and some egg masses will be destroyed or damaged. Therefore, it is estimated that the egg-mass density will be half the adult gypsy moth population.
[5]A further collapse of the host population could be expected during generation 6.

until the populations reach their peak. Based on this assumption, the rate of parasitism by solitary species will reflect the approximate ratio of adult parasites to adult hosts in the natural populations. If parasitism averages 20 percent, the ratio of adult parasites to adult hosts will be near 1:4.

In the model, the host population is estimated in terms of number of egg masses, since population density surveys are generally based on egg-mass data. According to Campbell (1981), female gypsy moths usually outnumber the males. However, some mortality of the females can be expected to occur before egg masses are deposited, and some mortality of the eggs will also occur. For convenience, therefore, the estimated ratio of adult parasites to adult gypsy moths is assumed to indicate the ratio of female parasites to egg masses. Thus, the ratio should be near 1:4 when parasitism averages 20 percent. If normal parasitism is higher or lower, the ratio will differ. For example, if parasitism averages 10 percent, the ratio of female parasites to egg masses would be near 1:9. If parasitism averages 25 percent, the ratio would be near 1:3, and so forth.

The life table data presented by Campbell, showing the number and fate of the various life stages of the gypsy moth when populations are in an outbreak mode, were used as a guide in developing the basic parameters for the host population during the first generation. The estimated number of adult *B. pratensis* coexisting with the gypsy moth population is based largely on the assumed rate of parasitism and the comparable survival hypothesis. As the two populations grow, density-dependent suppression forces increase. But they act with greater intensity against the gypsy moth population than on the parasite population about the time the host population begins to reach its peak. As a result, the ratio of adult female parasites to egg masses increases. This increase can account for the reported increases in rates of parasitism by *B. pratensis* during the collapse phase of gypsy moth populations. It should be noted, however, that the average number of host larvae parasitized per female parasite declines progressively despite a near-100-fold increase in the larval population from the low to the high density.

This inverse relationship suggests that there is more likely a negative rather than a positive correlation between host density and the average host parasitization ability of the female parasites during their lifetime.

Only the larger host larvae are able to ingest the parasite eggs (Godwin and Odell 1981). Therefore, the estimated number of the larger larvae (4th to 6th instar) produced per egg mass and their survival rates are key parameters. The egg-mass density reaches a peak of more than 5,000 per acre during generation 5. The population of large larvae reaches a maximum of nearly 300,000 per acre during generation 4 and then begins to collapse. Severe defoliation could be expected during generations 4 and 5. I consider the basic model to be a very realistic projection of the normal numerical relationship between the parasite and the gypsy moth.

General Discussion

As noted in chapter 2, the rate of parasitism among many parasite-host associations remains rather constant or within rather narrow limits, even though the host population may fluctuate widely from the normal low to the normal high level. Thus, the model is consistent with those observations. Such a parasitism trend will occur if the relative numbers of host larvae produced and larvae parasitized remain fairly constant and also if the relative survival rates of unparasitized host larvae and immature parasites remain essentially equal. As the host and parasite populations increase, their survival rates can be expected to decrease, and the actual number of host larvae produced per female and number parasitized per female parasite can be expected to decline. Godwin and Odell (1981) referred to data obtained by Semevsky (1973), which indicated no correlation between the gypsy moth densities and the proportion of the host pupae parasitized. The levels of parasitism observed by Semevsky averaged near 20 percent. Table 18 shows a rather constant rate of parasitism, near 20 percent, until the larval host population approaches the peak. Then, consistent with Godwin and Odell's observation, the rate of parasitism begins to increase as the larval host population begins to collapse.

Why does such an increase in parasitism occur? The inclination might be to assume that the increase is due to a positive response to the increase in host density. However, when I attempted to develop a model based on this assumption, it soon became apparent that unrealistic values would have to be assigned to the other parameters in order for the model to show both a rather constant rate of parasitism during the population growth phase and a substantial increase in the rate of parasitism during the collapse phase. Therefore, I concluded that during the period that the gypsy moth population rapidly approaches the peak and then begins to collapse, the parasitism trend is determined by relataive changes in the average number of host larvae surviving per egg mass and the average number of host larvae parasitized by the coexisting female parasites. My views on the manner that parasitism is achieved by *B. pratensis* in a natural environment may be summarized as follows:

> The ratio of stable, coexisting populations of *B. pratensis* and the gypsy moth is fairly constant. In the model, the ratio of female parasites to egg masses is assumed to be 1:4, which results in 20 percent parasitism of the large larvae. Conditions that favor an increase in the gypsy moth population also favor an increase in the parasite population; and the two populations tend to grow at a similar rate during an outbreak mode. The female parasites concentrate their eggs on foliage in larval habitats in response to kairomones produced by the feeding larvae. This parasitic response is suggested by Godwin and Odell (1981), Leonard (1981), and Odell and Godwin (1984). As the larval host population increases, the size of the habitats where larvae are present tends to increase proportionally.
>
> A comparable increase in the parasite population would then mean that the number of parasite eggs per unit of foliage in the larval habitats will tend to remain about the same. Thus, the probability that a parasite egg will be ingested remains about the same. The total number of parasite eggs deposited and the total number of host larvae present may increase manifold, but the ratio of leaves having parasite eggs to those that are free of parasite eggs will be similar. Hence the probability that a given individual larva will ingest parasite eggs remains fairly consistent as the populations grow. Therefore, the rate of parasitism will also remain reasonably constant.
>
> As the gypsy moth and the *B. pratensis* populations grow, density-dependent suppression forces intensify for both organisms and result in slower but comparable growth rates. However, about the time the host population peaks, the total mortality of the gypsy moth adults, eggs, and small larvae increases to a greater extent than that of the comparable life stages of *B. pratensis*. Therefore, the average number of large host larvae produced per egg mass declines more rapidly than the average number parasitized per female parasite. This causes a temporary increase in the rate of parasitism. The mortality of all of the immature larval stages of the parasite

population remains closely linked with the mortality of the large gypsy moth larvae and pupae. But during the total collapse phase, very few large larvae and pupae will survive. Therefore, very few parasites will complete development.

The coexistence model offers a plausible explanation for increases in the rate of parasitism that occur not only in the *B. pratensis*-gypsy moth association but also in other parasite-host associations that reach very high population levels and then collapse. The main point is that rates of parasitism may increase when host populations approach or reach their peak, but such an increase cannot be attributed to an increase in the host-finding ability of individual parasites in response to an increase in host density.

The significance of the model cannot be fully appreciated unless we relate the number of larvae parasitized by the coexisting female *B. pratensis* to the number of host larvae available to be parasitized. If the model is reasonably accurate as regards both the ratio of adult female parasites to egg masses for natural populations and the average host-parasitization capability of the *B. pratensis* females, it suggests that an egg-mass density as low as 20 per acre has little or no adverse influence on the average host-parasitization capability of the parasite. If this lack of adverse effect should hold at even lower egg-mass densities, it would have very important practical implications. It would mean that the effectiveness of a given number of released parasites would be inversely proportional to the host density. Thus, over a range of virtually all host densities, the lower the egg-mass density, the fewer the number of *B. pratenis* is required to achieve a given rate of parasitism. The egg deposition behavior of *B. pratensis* supports this assumption. The females may have no difficulty finding sites where host larvae are feeding, even when the egg-mass density had been as low as five, or perhaps even one, per acre. In such event, the number of parasite eggs deposited per unit of gypsy moth foliage in larval host habitats will be essentially the same whether the egg-mass density had been 50 per acre, 20 per acre, or as low as 1 per acre. Therefore, the probability that a parasite egg will be ingested would also be about the same. However, to achieve a 5:1 ratio of female parasites to egg masses, for example, only 5 female parasites per acre would be required when the egg-mass density averages 1 per acre, as compared with 250 females per acre when the egg-mass density averages 50 per acre. If the theoretical possibilities are real, the parasite augmentation technique should offer unusual opportunities to cope with low-density gypsy moth populations in an effective and ecologically acceptable manner. In the section to follow, an estimate is made of the influence of *B.*

pratensis releases in a gypsy moth ecosystem when the initial egg-mass density averages 20 per acre.

Influence of *Blepharipa pratensis* Releases

Suppression Model

An estimate was made of the impact of *B. pratensis* releases on the rates of parasitism and on the dynamics of a gypsy moth population like that shown in table 18. The egg-mass density averages 20 per acre during generation 1. The parameter values for the host population during generation 1 were used in calculating the effect of the parasite releases. The suppression model, shown in table 19, is based on the assumption that enough *B. pratensis* are released to achieve a 2.5:1 ratio of female parasites to egg masses. It should be noted that a ratio of 1:4 is estimated to result in 20 percent parasitism of the host larvae and pupae. Since a 2.5:1 ratio would mean a 10-fold increase in the parasite population, a high rate of parasitism could be expected. On the other hand, predation of the parasite adults and eggs is likely to increase. Therefore, for the augmented population, the average number of host larvae parasitized per female parasite is assumed to be 102 during generation 1, or 20 percent less than for a normal parasite population. This percentage reflects the rate of increase in predation due to the abnormally high parasite population.

Based on these assumptions, calculations were made of the effect of the parasite releases during year 1 and the effect of the naturally produced progeny during years that follow. The procedures followed in calculating the theoretical results presented in table 19 are largely self-explanatory. The calculations show that the ratio of parasite-host encounters to host larvae present would be 1.6:1 during generation 1. When allowance is made for the encounter of some hosts already parasitized, the level of parasitism would be approximately 80 percent. Because of the high intrinsic rate at which the host population increases, this level of parasitism would not cause a decline in the gypsy moth population. Theoretically, the egg-mass density would increase to an average of 40 per acre during year 2. Therefore, host density would not decline and neither should the host-finding capability of the parasites. However, the ratio of adult female parasites to egg masses would theoretically increase from 2.5:1 to 4:1 because of the comparable survival theory. The average number of host larvae parasitized per female parasite during year 2 is assumed to be reduced by 10 percent because of the further increase in the adult parasite and

Table 19

Suppression model depicting the estimated influence of *Blepharipa pratensis* releases in a gypsy moth ecoystem if enough *B. pratensis* are released to result in a 2.5:1 ratio of adult parasites to egg masses during generation 1 of the host population in an outbreak mode

Parameter (1 acre)	Generation			
	1	2	3	[1]4
Gypsy moth egg masses	20	40	40	6
Large larvae per mass[2]	160	160	160	200
Total large larvae	3,200	6,400	6,400	1,200
Female *B. pratensis*	50	160	324	284
Ratio, adult female parasites to egg masses	2.5:1	4:1	8.1:1	47:1
Host larvae parasitized per female[3]	102	92	82	—
Total parasite-host encounters	5,100	14,720	26,568	—
Ratio, parasite-host encounters to host larvae present[4]	1.6:1	2.3:1	4.2:1	—
Percent larvae parasitized	80	90	98.5	99
Host larvae parasitized, number	2,560	5,760	6,304	1,188
Host larvae not parasitized	640	640	96	12
Survival rate of unparasitized larvae to adults (percent)[2]	12.5	12.5	12.5	15
Gypsy moths emerging	80	80	12	2
Egg masses, next year[5]	40	40	6	1
Survival rate of immature parasites to adults (percent)[6]	12.5	11.25	9	4
Adult *B. pratensis*, next year	320	648	568	48

[1]Dashes indicate that no estimates are made for generation 4. However as explained in the text, the rate of parasitism should be the assumed maximum of 99 percent.
[2]In view of the slight changes in egg-mass density, no adjustments are made in the number of host larvae produced per egg mass and their survival rate until generation 4.
[3]In view of the increasing parasite population, the average number of hosts parasitized per female is assumed to decline.
[4]See table 1 to calculate the rates of parasitism and escape from parasitism by using the ratio of parasite-host encounters to hosts present.
[5]The ratio of sexes may favor females. However, considerable mortality of the females may occur before eggs are deposited, and some egg masses may be lost. Therefore, the egg-mass density is assumed to be 50 percent of the number of moths emerging.
[6]According to Godwin and Odell (1981), excessive reparasitizations cause some premature mortality of the superparasitized host larvae. As the rate of parasitism increases, the probability of superparasitism increases. The estimated survival rates reflect such increases in mortality.

parasite egg populations. This negative factor would not equal the large increase in the ratio of parasites to egg masses. Therefore, the ratio of parasite-host encounters to hosts present would increase from 1.6:1 to 2.3:1, an increase of almost 50 percent. This new ratio would theoretically increase the rate of parasitism to 90 percent.

According to Godwin and Odell (1981), excessive reparasitization causes the premature mortality of some larvae. Thus, it would also cause the premature mortality of the parasites within those larvae. When parasitism averages 90 percent, the host larvae will be parasitized an average of about 2.5 times. Godwin and Odell also observed, however, that some superparasitized larvae produce more than one adult progeny. Thus, it is questionable that the survival rate of the immature parasites will decrease when parasitism is 90 percent. To be conservative, however, a reduction of about 10 percent is assumed.

Since the gypsy moth population in this case is assumed to have the potential to increase 10-fold, 90 percent mortality due to parasitism of the larvae, in addition to the mortality resulting from normal hazards, would merely stabilize the population. Therefore, the egg-mass density during generation 3 is assumed to remain at 40 per acre; and host density should again have no significant influence on the average host parasitization capability of the female parasites. But the parasite population in generation 3 is two times that in generation 2; and this increase is estimated to reduce the average number of larvae parasitized by 10 percent. Such a reduction would be a relatively minor negative factor, however, because the ratio of parasites to egg masses has more than doubled to 8.1:1. Therefore, the rate of parasitism during generation 3 is estimated to increase to 98.5 percent.

The estimated influence during generation 4 becomes more speculative because several major changes would have occurred by that time. The egg-mass density would average six per acre, a decline of about 70 percent from the density in generation 1. However, this low density may have no significant adverse effect on the host-habitat-finding capability of the female parasites. On the other hand, the high rate of parasitism during generation 3 would mean that the larvae would be parasitized an average of about four times; and I arbitrarily assume that this rate of superparasitization would reduce the survival rate of the immature parasites by about 28 percent. Nevertheless, this negative effect would be minor compared with the theoretical 5.8-fold increase in the ratio of adult parasites to egg masses from generation 3 to generation 4, or a ratio of 47:1. The parasite population would have declined slightly, so no further increase in the rate of predation would be expected. Without presenting details here, I estimate that a gypsy moth larva will ingest an average of about 0.33 egg when the ratio of female parasites to egg masses is 1:4. On this basis, a larva would ingest an average of about 60 eggs when the ratio is 47:1. In any event, theoretical parasitism during generation 4 is assumed to be 99 percent. Virtual collapse of the gypsy moth population should be the result.

Discussion of the Model

If the suppression model accurately depicts the influence of *B. pratensis* releases on rates of parasitism when a gypsy moth population with an egg-mass density of 20 per acre is in an outbreak mode, the parasite augmentation technique would be remarkably effective. Theoretically, a parasite-to-egg-mass ratio that is initially 2.5:1 would increase progressively, resulting in progressively higher rates of parasitism until the gypsy moth population completely collapses.

Obviously, however, a number of questions can be raised regarding the theoretical results. To my knowledge rates of parasitism that occur in natural populations at various egg-mass densities have not been determined. Also, it might be argued that the maximum level of parasitization may be well below 99 percent, regardless of the parasite-to-egg-mass ratio or the number of parasite eggs ingested by the larvae. A number of aspects of the biology and behavior of *B. pratensis* as determined by Odell and Godwin (1984) and Godwin and Odell (1984) were not considered in the theoretical appraisal.

Despite the uncertainties, I feel that the basic assumptions on which the model is based are reasonable. I question whether *B. pratensis* could survive unless it has the capability of parasitizing the required quota of hosts when the egg-mass density is very low, as it is when populations are in a stable endemic mode. If a model is developed that depicts a stable endemic population averaging 20 egg masses per acre and it is assumed that the female parasites will parasitize an average of only one-half as many larvae as are produced per egg mass, the parasite population would virtually disappear within about 6 years; and it will disappear in about 3 years if the endemic population averages 5 egg masses per acre.

If a 1:4 ratio of parasites to egg masses results in 20 percent parasitism, it would seem reasonable to assume that an increase in the ratio to 2:5:1, by artificial means, would result in parasitism on the order of 80 percent. Similar survival rates of the immature parasites and unparasitized hosts to the respective adult stages would seem logical. Such an initial high rate of parasitism and the theory of comparable survival rates for the immature parasites and unparasitized larvae to the respective adult stages account for the progressive increase in the ratio of parasites to egg masses and the progressive increase in the rate of parasitism in the subsequent generations. A declining host density would not be a factor in this case until generation 4. In theory, the ratio of parasites to hosts will steadily increase and exceed by up to 100-fold the maximum ratios that could develop in self-perpetuating populations.

If, indeed, it is confirmed that the primary factor governing the rate of parasitism by *B. pratensis* is the ratio of adult parasites to egg masses coexisting in the populations, even when the gypsy moth density averages only a few egg masses per acre, the parasite augmentation technique would offer unusual opportunities for suppressing or regulating gypsy moth populations. The technique, however, would have to be employed when the gypsy moth population is low. It would be impractical for controlling the pest after populations have reached threatening damage levels. In fact, I question the practicality of using the technique once egg-mass densities have reached an average of 50 to 100 per acre. But at this time, there may not be a more efficient method for preventing outbreaks if suppressive measures are taken before populations reach threatening damage levels.

If an initial parasite-to-egg-mass ratio of 2.5:1 or even 5:1 is high enough to result in high rates of parasitism when egg-mass densities average 20 per acre or less, the prospects seem excellent that the parasite release method will provide a practical means of maintaining low gypsy moth populations, especially when integrated with appropriate autocidal techniques. In the section to follow, appraisals will be made of the potential of the augmentation technique under different circumstances; and the appraisals will be based on the theoretical results that have been presented.

Possible Circumstances for Releasing *Blepharipa pratensis*

The suppression model, as noted, indicates that for high rates of parasitism, the parasite-to-egg-mass ratio must be on the order of 2.5:1 and the egg-mass density must be low—about 20 per acre or lower. It may, therefore, be desirable to consider the circumstances under which the parasite release technique might be useful, the probable costs, and what the limitations of the technique are likely to be. Four potential uses will be considered: (1) prevent outbreaks in urban and recreational areas, (2) delay the spread of populations into new areas, (3) prevent periodic outbreaks in long-established areas, and (4) eradicate new infestations by integrating the technique with autocidal techniques.

Mass-rearing methods for the gypsy moth are already well advanced (Bell et al. 1981). Some progress has also been made in rearing the parasite *B. pratensis* (Odell and Godwin 1979, 1981). There is reason to assume that the mass-production of the parasite will be operationally feasible and at costs that will allow the parasite to be used in augmentation releases for suppressing low-density populations of the gypsy moth. I make the assumption that the eventual cost of rearing the parasite in large numbers will be about $15 per 1,000 of both sexes. A suitable alternate host, or perhaps in vitro rearing methods, might eventually make it possible to rear the parasite at lower costs.

As previously implied, the maximum average egg-mass densities that could be cost effectively managed by the release of *B. pratensis* are likely to be on the order of 50 to 100 per acre. If a 2.5:1 ratio of female parasites to egg masses is required and if the egg-mass density averages 100 per acre, 500 parasites of both sexes would have to be released. Assuming a rearing cost of $15 per 1,000 for both sexes, the cost per acre for parasites alone would be nearly $7.50. On the other hand, if preventive measures could be taken when average egg-mass densities are on the order of 5 to 20 per acre and the efficiency of the parasite is not significantly reduced at such low densities, no other control procedure known at this time is likely to be less costly. Also, of special significance is the fact that the suppression model indicates that the parasite augmentation technique would result in continuing control for several years due to the parasite progeny produced, whereas alternative methods may have to be used more frequently.

Preventing New Infestations in Urban and Recreational Areas

When chemicals are used to control gypsy moths in residential and recreational areas, there is probably no greater concern than the potential health and environmental hazards that might be created by the chemicals. The release of a gypsy moth-specific parasite as a means of suppression should completely allay any concern about potential health or environmental hazards.

The factor that is most likely to limit the effectiveness of the parasite releases would be the excessive dispersal of the released parasites and/or their progeny out of the areas requiring protection. Considerable movement of the parasites within the release area would be advantageous if the release area is large and the parasites tend to concentrate in habitats where infestations are high. The theoretical results previously discussed are based on the assumption that the released parasites and their progeny exist in closed ecosystems. However, if releases are made in small unisolated areas, excessive movement of the parasites would greatly reduce the effectiveness of the technique. If half of the released parasites and their progeny move out of areas where protection is needed, calculations will show that the efficiency will be only about 25 percent of the theoretical.

Nevertheless, the potential of the technique is so promising, both from economic and ecological standpoints, that it should be thoroughly investigated. If suppressive measures were undertaken when the densities average 5 to 10 egg masses per acre, the technique might be the most practical and acceptable to employ even if releases were necessary at the rate of 100 parasites per acre every 2 or 3 years. If the estimate of $15 per 1,000 parasites is realistic, the cost of 100 parasites would be $1.50 per acre. Obviously, such a program would have to be supported by thorough surveys as a guide to the proper timing and distribution of the parasite releases.

Delaying Gypsy Moth Spread Into New Areas

In areas not yet infested but vulnerable to infestation by the gypsy moth, a major problem is to detect infestations and to eliminate or suppress them by practical procedures before they become large and extensive. In theory, the elimination or suppression could be accomplished by the release of highly mobile parasites that have the capability of finding the host habitats, even though the habitats may be small and widely scattered. Of the control measures now available, none has this intrinsic capability to seek out the target pest. In areas that are within 25 miles from the leading edge of gypsy moth populations that are well established but held to a low level by some control program, the egg-mass densities may average less than 1 egg mass per acre, even though the densities at some localized sites may be 10 to 50 egg masses or more per acre. The release of an average of 25 parasites of both sexes per acre each year in such 25-mile-wide barrier areas could be expected to result in an overall female-parasite-to-egg-mass ratio that would ensure effective control. If the parasites have the capability of readily locating small isolated infestations even though the infestations may be widely scattered, such a release rate may result in 90 percent parasitism or more, a result I believe would be necessary to prevent spreading populations from increasing. Assume that 10 small infestations exist within a 1-square-mile area, that the egg masses total 500, and that parasites of both sexes are released at the rate of 25 per acre (= 16,000 parasites of both sexes, or 8,000 females, per square mile). The resulting overall ratio of female parasites to egg masses would be 16:1. The parasites would probably disperse widely; but if even as few as 25 percent of the females were successful in locating the infested habitats, the effective ratio would still be 4:1. Theoretically, the rate of parasitism would be about 90 percent. Quite possibly all the parasites except those in the border areas would be more than 25 percent efficient in locating the infested habitats. While the movement of most of the parasites out of the barrier zone would be likely, and undesirable, the potential of the technique is so great that it deserves consideration and appropriate experimentation.

The release of irradiated male moths to serve as a barrier to spread is also being considered. However, if a 4:1 ratio of female parasites to egg masses results in 90 percent larval parasitism, a 9:1 ratio of fully competitive sterile males to fertile males would be required in order that the sterile moths be as suppressive as the parasites. Irradiated moths are not likely to be fully competitive. Also the sterile moths would doubtless have to be released somewhat proportionally to the distribution of the natural population, and timing of the releases would be critical. Locating the infested sites and determining when adult emergence occurs would be major difficulties. The greatest advantage of the parasites may be their ability to readily find the infested sites on their own and also their effectiveness over more than one generation. In my opinion, the probability that released parasites will find scattered larval habitats is much greater than the probability that released males will find and mate with the native female moths in widely scattered infestations before the native females can mate with native males that emerge in the same vicinity. The cost of rearing and releasing parasites is not likely to exceed the cost of rearing and releasing sterile males. For these reasons the parasite release technique may be many times more effective and considerably less costly per unit area than the autocidal technique when costs are averaged over several years.

Preventing Periodic Outbreaks of Established Populations

The feasibility of releasing *B. pratensis* to prevent periodic outbreaks of long-established gypsy moth populations would depend largely on the answers to two questions: (1) What is the average size of gypsy moth populations when they are in a stable, low-density mode? (2) What is the host-finding capability of the female parasites when the larval populations are low and widely scattered, as they are likely to be in low endemic populations? Definitive answers to these questions are not yet available, so thorough ecological investigations designed to quantitatively answer these questions would be highly desirable.

But if the average egg-mass densities under normal stable conditions should be as low as five per acre or lower in a large ecosystem (most of the population concentrated, no doubt, in only parts of the total ecosystem) and if *B. pratensis* is found to be capable of surviving under such circumstances (thereby demonstrating its effectiveness in finding hosts), this parasite could be suitable for use in augmentative releases to prevent periodic gypsy moth outbreaks. Although excessive movement of the released parasites and their progeny was cited earlier as a probable limitation to suppressing gypsy moths in small unisolated areas, such movement could be of great advantage in suppressing the moths on a large-area basis.

At egg-mass densities averaging as low as five per acre or lower, the release of relatively few parasites might raise the parasite-to-host ratios high enough to result in high rates of parasitism. The introduction of a new hazard to survival when a gypsy moth population is already under considerable environmental stress could have a highly adverse impact on the dynamics of the population. Perhaps the release of as few as 25 to 50 parasites per acre at 5-year intervals could prevent periodic outbreaks of the gypsy moth in critical areas. Almost certainly, however, such releases would not be effective unless made on an areawide basis, perhaps involving a minimum of several thousand square miles.

Integrating Augmentation and Autocidal Techniques To Eradicate New, Isolated Gypsy Moth Infestations

The characteristics of the parasite augmentation and autocidal techniques were discussed in some detail in chapter 3. While the two techniques have several characteristics that are similar, they also have characteristics that are different. When the two techniques are appropriately integrated, their suppressive actions are synergistic, and the end result is a suppressive action of extraordinary intensity. The most important limitation of both techniques is that the higher the pest density, the greater the number of insects that must be released to achieve the required ratio of released insects to the native pest population. Thus, as previously noted, the feasibility of using either technique will be restricted to pest populations existing at rather low levels. The integration of parasite and sterile insect releases could be advantageous for suppressing gypsy moth populations in residential areas and for delaying the spread of infestation, as discussed earlier. But in my view, the two techniques employed concurrently may be particularly useful for the eradication of newly discovered infestations.

It is usually urgent that such infestations be eliminated as quickly as possible. The number of gypsy moths present is generally very low, and the cost involved in the elimination is usually a relatively minor consideration. Chemical insecticides can be effective and practical for eliminating such infestations. But, as previously noted, there may be strong opposition to their use for fear of health or environmental hazards. Since gypsy moth-dependent parasites and sterile insects are target-pest specific, their use would preclude questions of hazards.

Why the release of *B. pratensis* and sterile males can be advantageous in eliminating newly discovered gypsy moth infestations from isolated areas will be described rather briefly. The release of "partly sterile males" (males receiving a dosage of irradiation that would cause a high percentage of their F_1 progeny to be sterile) would likely be more effective than the release of fully sterile males. However, a complex model would be required to describe how the two techniques would interact to provide the advantage. A unique method involving the distribution of F_1 egg masses from partly sterile males mated to normal females may also be used for suppression. LaChance (1985) reviewed the research that has been conducted on these sterility techniques for gypsy moth suppression. In principle, the advantages would be similar if the parasite augmentation technique were integrated with any of the sterility techniques, but the advantage can be described more readily based on sterile male releases.

The models shown in table 20 indicate the theoretical advantage of releasing parasites and sterile males concurrently during the first year over releasing sterile males only. The models are somewhat self-explanatory, but brief comments will be made on some of the assumptions on which the models are based.

When discovered, the numbers of egg masses are assumed to total 1,000. Most of the infestation is likely to be concentrated in a rather small area, but because of small-larvae movement, the

Table 20
Two models depicting the relative efficiencies of a sterile-male-release technique (model I) and a combination technique involving *Blepharipa pratensis* and sterile male releases (model II) for eradicating newly discovered gypsy moth infestations (see text for details)

Parameter	Model I, 2 million sterile males released yearly				Model II, 100,000 female *B. pratensis and* and100,000 sterile males released in year 1, and 100,000 sterile males released in year 2	
	Year[1]				Year	
	1	2	3	4	1	2
Egg masses	1,000	500	125	8	1,000	25
Expected adults per egg mass	20	20	20	20	20	20
Expected adult population	20,000	10,000	2,500	160	20,000	500
Female *B. pratensis*	NA	NA	NA	NA	100,000	[2]5,000
Overall ratio, *B. pratensis* to egg masses	NA	NA	NA	NA	100:1	200:1
Effective ratio, *B. pratensis* to egg masses	NA	NA	NA	NA	[3]5:1	[3]10:1
Host larvae parasitized, percent	NA	NA	NA	NA	95	99
Adult gypsy moths emerging	20,000	10,000	2,500	160	1,000	5
Sterile males released	2,000,000	2,000,000	2,000,000	2,000,000	100,000	100,000
Overall ratio, sterile males to fertile males	200:1	400:1	1,600:1	25,000:1	200:1	20,000:1
Estimated effective ratio, sterile males to fertile males[4]	20:1	40:1	160:1	2,500:1	20:1	2,000:1
Females mating with fertile males	500	125	8	0	25	0
Egg masses, next year	500	125	8	0	25	0

[1]NA = not applicable.
[2]The estimated adult parasite progeny that survive from year 1.
[3]Based on the assumption that only 5 percent of the female *B. pratensis* will concentrate in larval host habitats.
[4]Based on the assumption that only 10 percent of the released moths will be physiologically and spatially competitive.

adults that develop from the egg masses may reproduce in several thousand acres. Conditions for reproduction are often favorable in new areas of spread, and the population may increase by 10-fold per year if not suppressed. If sterile males alone are released, the adult population produced from the 1,000 egg masses would number 10,000 males and 10,000 females. Because of the difficulty of distributing the moths proportionally to the distribution of the native adults, moth movement, and irradiation damage, the released moths are estimated to be only 10 percent competitive. To ensure a decline in the population, 2 million sterile males are proposed to be released each year until reproduction ceases. The effects of the releases are shown in table 20.

As compared with only sterile male releases, a combination of sterile male and *B. pratensis* releases during year 1 and then only sterile male releases during year 2 could be expected to be more quickly and greatly suppressive. The model shown is based on the release of 100,000 female *B. pratensis*. An overall ratio of 100 female parasites to 1 egg mass would be expected; however, because of the distribution pattern of the host larvae, 95 percent of the parasites are assumed to disperse out of the infested area. The remaining 5 percent are assumed to contribute to control and would result in a 5:1 ratio of female parasites to egg masses. Based on the estimated efficiency of the parasites, a ratio of 5:1 is assumed to result in 95 percent parasitism of the large larvae. If such is the result, the adult gypsy moth population produced from the egg masses should be about 1,000 of both sexes. Then, the release of 100,000 sterile males would mean an overall sterile-to-fertile ratio of 200:1 and an effective ratio of 20:1 based on 10 percent competitiveness. The combined effect of 95 percent parasitism and 95 percent inhibition of reproduction by the adults during the same generation would theoretically reduce the number of egg masses during year 2 to 25. This assumes a 10-fold increase rate for an uncontrolled population.

The parasite progeny produced during year 1 should be adequate to result in a 200:1 ratio of female parasites to egg masses during year 2. But if only 5 percent of the females contribute to control, the effective ratio would be 10:1. In theory, this ratio should be adequate to almost eliminate the population, but it remains to be determined whether parasite releases can lead to eradication. If 100,000 sterile males are again released, the overall sterile-to-fertile ratio would be on the order of 20,000:1, with an effective ratio of about 2,000:1. No successful reproduction would be expected during year 2.

In theory, the release of the parasites would enhance the effectiveness of the sterile male release by more than 20-fold. The sterile males would also enhance the effectiveness of the parasites during year 2. In terms of relative costs, if the costs of rearing the female parasites and sterile males for release amount to $30 per 1,000, the integrated approach would cost about $9,000 for the insects. In contrast, if 8 million sterile males were required for eradication, the cost for the sterile insects would be $240,000. We might also consider costs to eradicate a new, isolated infestation by larvicide spraying. If the egg-mass density has reached 1,000 when discovered, we might assume some larvae in a 25-square-mile area. Assuming 15,000 acres of gypsy moth habitats and three sprays during a period of 3 years at $5 per acre for each spray, the cost of eradication would be $225,000.

This analysis is based on assumptions that may be speculative in several respects. They may also be conservative in other respects. Certainly, however, the basic principles are biologically sound. The mechanisms of suppression involving the use of parasites and sterile males could add a new dimension to future insect-pest-population suppression and eradication technologies that would have great practical significance. Moreover, both techniques would be target-pest specific.

Effects of Random and Selective Deposition of *Blepharipa pratensis* Eggs on Rate of Parasitism of the Gypsy Moth

The capability of parasites to find and parasitize their necessary quota of hosts even though host populations are often abnormally low has been proposed as one of the most fundamental behavioral characteristics that has ensured parasite survival for many years. Also, as previously discussed, this capability has important practical implications in the parasite augmentation technique. There seems to be little doubt that most parasites could not survive for long if they relied on random searches for hosts, especially when the hosts are scarce and widely distributed. It is difficult, however, to quantitate the advantage of host detection through use of kairomones and other chemical signals over host detection by random discovery. Therefore, we must again rely on deductive procedures to gain some idea of the magnitude of the advantage, if any.

Two models were developed to help quantify such an advantage. They are based on the assumption that for depositing their eggs, *B. pratensis* females will select foliage in random sites or only foliage that is in sites harboring feeding gypsy moth larvae.

The models are shown in table 21 and are largely self-explanatory, but some of the basic assumptions will be discussed briefly. The gypsy moth populations are assumed to be in a stable mode and average 20 egg masses per acre. The life-table data for a stable, low-density population presented by Campbell (1981) were used as a guide in estimating values for parameters relating to the gypsy moth populations. Basic information obtained by Zecevic (1958), Braham and Witter (1978), and Valentine (1983) and unpublished data kindly made available by Ralph Webb (U.S. Department of Agriculture) were helpful in estimating both the amount of foliage consumed by mature larvae and the amount of foliage present per acre of typical gypsy moth habitats. The *B. pratensis* population either exists or is added to the gypsy moth environment at the rate of five females per acre. Thus, the ratio of female parasites to egg masses would be 1:4. Based on the average number of eggs deposited by the female parasites and the amount of foliage consumed by the gypsy moth larvae by the time they are ready to pupate, it is estimated that the ratio of leaves with parasite eggs to leaves without eggs in gypsy moth larval habitats would have to be about 1:18 in order for the rate of parasitism to be 20 percent. In such case, each larva would have ingested an average of about 0.33 egg.

For the population that is assumed to deposit eggs in random sites, all of the basic parameters relating to the host, the parasite, and the host plants remain constant, but the parasite eggs are assumed to be distributed strictly at random sites. During the first year, the ratio of leaves with parasite eggs to leaves without eggs is estimated to be 1:200. At this ratio, so few eggs would be ingested by the feeding larvae that the rate of parasitism in year 1 would be only about 1.8 percent. Allowing for natural mortality, the parasite population would be so low during year 2 that the ratio of leaves with and without eggs would decrease by more than 10-fold. This decrease would further decrease the probability that the feeding larvae will ingest parasite eggs. The rate of parasitism during year 2 is estimated to decline to about 0.17 percent. Although calculations are not shown, the probability that even one host larvae per acre would ingest a parasite egg during year 3 would be virtually nil. In effect, the parasite population would disappear within about 3 years.

In contrast to egg deposition on random foliage, egg deposition only in habitats harboring feeding host larvae would enable *B. pratensis* to maintain a steady density population. At the host density assumed, the probability of successful parasitization is increased by approximately 11-fold because of the host-guidance mechanism.

Table 21
Theoretical relative efficiencies with which gypsy moth larvae ingest *Blepharipa pratensis* eggs when the eggs are deposited on leaves in random sites (random distribution) and when the eggs are deposited only on leaves within habitats harboring the feeding host larvae (selective distribution)

Parameter (100 acres)	Random distribution		Selective distribution	
	Year 1	Year 2	Year 1	Year 2
Gypsy moth egg masses	2,000	2,474	2,000	2,000
Mature larvae per egg mass	12	12	12	12
Total mature larvae	24,000	29,688	24,000	24,000
B. pratensis females	500	46	500	500
Ratio, *B. pratensis* females to egg masses	1:4	1:54	1:4	1:4
Eggs deposited per female parasite	1,500	1,500	1,500	1,500
Total eggs deposited	750,000	69,000	750,000	750,000
Total host leaves (millions)	150	150	150	150
Ratio, leaves with eggs to leaves without eggs[1]	1:200	1:2,174	1:13	1:18
Average number leaves consumed by larvae	6	6	6	6
Total leaves consumed	144,000	178,128	144,000	144,000
Total parasite eggs ingested	720	82	8,000	8,000
Percent ingested eggs resulting in parasitism[2]	60	60	60	60
Number parasitized larvae	432	49	4,800	4,800
Percent larvae parasitized	1.8	.17	20	20
Larvae not parasitized	23,568	29,639	19,200	19,200
Percent survival to adults	21	21	21	21
Adult gypsy moths next year (both sexes)	4,949	6,224	4,032	4,032
Gypsy moth egg masses next year[3]	2,474	3,112	ca. 2,000	ca. 2,000
Adult *B. pratensis* next year (both sexes)	91	10	ca. 1,000	ca. 1,000

[1]The ratio for random distribution of eggs is calculated. The ratio for selective distribution would have to be 1:18 in order to result in 20 percent parasitism, a basic premise.
[2]Based on data by Godwin and Odell (1981).
[3]The ratio of females to males often favors the females. However, because of some mortality of the females before egg deposition and also some mortality of eggs, the egg-mass population is assumed to be 50 percent of the moth emergence.

Although models will not be shown, the advantage of the host-guidance factor over absence of the factor depends on the host density. Calculations indicate that in the absence of this factor, a *B. pratensis* population could maintain a viable population only if the egg-mass density of the gypsy moth population is at least about 300 per acre. This density would suggest that the advantage of selective over random egg distribution is on the order of 15-fold when the egg-mass density averages 20 per acre. If data were now available on the minimum average egg-mass density necessary for the two populations to coexist successfully, a more precise estimate could be made of the advantage of selective over random egg distribution. The models are based on the assumption that this minimum density is 20 egg masses per acre. If the density must be higher, the advantage would be less. If the populations can coexist in a stable state at a lower host density, the advantage would be even greater than estimated. The results of this analysis are generally consistent with a working hypothesis adopted for this investigation, namely, that in host searching, parasites devote 10-fold as much time in host habitats as they do in sites that have no hosts.

The theoretical calculations in this section might seem to be largely a matter of academic interest. But such is not the case. While the parameter values used may not be highly accurate, the results are nevertheless consistent with the findings, made by parasite behaviorists, that biological agents possess highly developed host-guidance mechanisms. The results are also consistent with theoretical estimates by other procedures. The results provide some quantitative indication of the advantage of highly developed host-guidance mechanisms when host densities are low. From a practical standpoint, the theoretical estimates have special significance. They show that for maximum efficiency, parasite releases should be made when the host populations are low. If it can be shown that most species have the capability of parasitizing enough hosts to maintain viable populations when the host densities are very low, the parasite augmentation technique offers almost unlimited potential for suppressing and maintaining pest-host populations at levels that are not economically damaging.

Effect of Parasite Releases on Parasite Ingestion Rate and Parasitism Rate of the Gypsy Moth

Several parameter values used to estimate the advantage of selective deposition of *B. pratensis* eggs in gypsy moth larval habitats, as discussed for the model shown in table 21, can also be used to estimate the number of eggs likely to be ingested by the feeding larvae when the parasite population is artificially increased. Because of this selective deposition, the egg density per unit of foliage is assumed to remain reasonably constant for a given ratio of parasites to egg masses, even though the actual numbers of adult parasites and egg masses may vary. The estimates for the key parameters are as follows: (1) A 1:4 ratio of female parasites to egg masses will result in a parasite egg density of 1 per 18 average-size leaves in the larval habitats; (2) larvae that complete development will have consumed 6 leaves; and (3) among host larvae ingesting 1 egg each, 60 percent will become parasitized. This last estimate is considered consistent with data obtained by Godwin and Odell (1984). Based on these estimates, a 1:4 ratio of parasites to egg masses would result in the ingestion of 0.33 egg per larva and 20 percent parasitism in the larval host population.

The theories previously discussed are relevant to estimating the rate at which the feeding larvae ingest parasite eggs. Based on these theories, (1) *B. pratensis* females are assumed to respond to kairomone signals produced by feeding host larvae and to deposit their normal quota of eggs on foliage in the larval habitats, even though the average egg-mass density is as low as five per acre or lower and (2) the rates of survival of the immature parasites (ingested as eggs by host larvae) and unparasitized host larvae to the respective adult stages tend to be equal. Reparasitization is assumed to have no significant negative effect on the survival rate of the ingested parasites unless the host larvae are reparasitized several times.

Together, these parameter values and assumptions allow us to estimate not only the average numbers of parasite eggs that will be ingested by the larvae but also the rates of parasitism that will result in successive years following the release of *B. pratensis* in a gypsy moth ecosystem.

It is assumed that 25 *B. pratensis* of both sexes will be released per acre when the egg-mass density averages 5 per acre. Thus, the ratio of female *B. pratensis* to egg masses will be 12.5:5, or 2.5:1. While the distribution of the egg masses and the resulting gypsy moth larvae may be highly variable per unit area, this variability is not a factor, because of the assumption that the parasites have the capability of readily finding the host habitats and depositing their normal quota of eggs. If a 1:4 ratio of female parasites to egg masses results in an average of 1 egg deposited per 18 leaves, a ratio of 2.5:1 should result in an average of 1 egg per 1.8 leaves.

But because of the abnormally high adult parasite and parasite egg populations, the average is arbitrarily assumed to be 20 percent lower, or 1 egg per 2.2 leaves. Therefore, the larvae would ingest an average of 2.7 eggs (6 ÷ 2.2). Egg ingestion by the larvae is assumed to be random. A rather complex formula would be required to determine the proportion of the larvae that would ingest 0, 1, 2, 3, or more eggs. However, based on the data in table 1, which would be applicable in this case, the probability is that about 7 percent of the larvae will not ingest any parasite eggs when the ingestion rate averages 2.7. If 7 percent escape parasitism when the average is 2.7, an increasing percentage of larvae would ingest 1, 2, and 3 eggs, and a decreasing percentage would ingest 4, 5, and 6 eggs. If 40 percent of the larvae ingesting 1 egg escape parasitism, about 16 and 6 percent, respectively, of the larvae ingesting 2 and 3 eggs would escape parasitism.

The rate of parasitism is estimated to be approximately 83 percent when the ratio of parasites to egg masses is 2.5:1. This percentage would be near the 80-percent rate of parasitism estimated for a 2.5:1 ratio of parasites to egg masses by the procedure described for the model shown in table 19. If parasitism averages 83 percent during year 1, the ratio of female *B. pratensis* to egg masses should be approximately 5:1 during year 2 (83 ÷ 17 = 4.9). However, because of the higher parasite density, the effective ratio is arbitrarily assumed to be 4:1. If a ratio of 2.5:1 results in the ingestion of an average of 2.7 eggs, a ratio of 4:1 should result in the ingestion of about 4.3 eggs. In such event, parasitism should exceed 95 percent during year 2. The ratio of adult parasites to egg masses would then be about 19:1 during generation 3, provided the survival rates of the developing parasites and unparasitized host larvae to the respective adult stages continue to be equal. If the survival rate of the ingested parasites to the adult stage is reduced by 50 percent because of severe superparasitism, the effective ratio would still be high, near 10:1, and the parasite egg density on the foliage would increase to a level that would result in the ingestion of 8–10 eggs per larva.

By year 3, the gypsy moth population may be reduced to nearly zero. We have no knowledge about the effects of such a low population on the host-habitat-finding capability of the parasites. Biologists could readily determine, under laboratory conditions, the rates of parasitism that will result when the larvae ingest different numbers of eggs. They could also determine the survival rates associated with the numbers of eggs ingested. However, they may have difficulty in determining the influence of abnormally low host densities on the host-finding capability of the female parasites.

The purpose of estimating the rate of parasitism on the basis of the number of parasite eggs that are likely to be ingested by the host larvae was to compare the results with those obtained on the basis of the ratio of adult parasites to egg masses. The theoretical results by both methods are essentially the same.

9 Regulating Populations of *Heliothis* Species

Overview

Heliothis zea and *H. virescens*, as a complex, are the most costly agricultural insect pests in the United States. They attack a wide range of crops and wild host plants and are distributed virtually nationwide. In view of their abundance, diversity of host plants, and long flight range, the feasibility of rigidly regulating total populations by any means may seem beyond technical and practical realization. Without doubt, any programs developed for suppressing *Heliothis* populations would have to be of unprecedented scope—larger and more demanding in their execution than the screwworm programs (Graham 1985). Complex technical, financial, and operational problems would have to be solved. However, if the suppression of *Heliothis* populations could be realized cost effectively by ecologically acceptable procedures, the rewards would be great. It would prevent agricultural losses that are estimated to exceed $1 billion annually (U.S. Department of Agriculture 1976a). It would obviate the need for ecologically disruptive insecticides that now provide the primary means of controlling the pests after they have reached economically damaging levels.

Much of the Nation's research on insects during the past 25 years has been focused on the *Heliothis* species. These pests are of worldwide importance (King and Jackson 1989). Scientists engaged in the research have obtained much information that remains to be put into practical applications. New information has been obtained on the biology, behavior, and dynamics of *Heliothis* populations. Research has been conducted on parasites, predators, and microbial agents that contribute to control (Johnson et al. 1986). The search for host-plant resistance has continued on the principal crops (Maxwell and Jennings 1980). Chemists have identified and synthesized the sex pheromones the two species use for sexual communication (Klun et al. 1979). These attractants enable the development of highly sensitive detection methods and, thus, can contribute to more successful control applications. Outstanding progress has been made on rearing methods for both species and on some of the associated parasites and predators (King and Leppla 1984). New genetic manipulations that are potentially more effective than the conventional sterility techniques have been discovered (North 1975, Knipling and Klassen 1976, Laster et al. 1976, LaChance 1985).

Unfortunately, however, most of the advances on methods of control may contribute relatively little to more effective, more economical, or more acceptable solutions to the *Heliothis* problem, unless the general approach to *Heliothis* management is changed. The prospects seem dim for making practical use of genetic techniques, sex attractants, and highly mobile biological control agents for practical control of *Heliothis* on a crop-to-crop basis by individual growers after the pest populations have reached damaging levels. On the other hand, as will be discussed in subsequent sections of this report, such techniques may prove to be practical for maintaining populations below economically damaging levels if directed against total populations early in the season, when the populations are low and restricted in distribution.

It has long been my conviction that a satisfactory solution for these wide-ranging and dynamic pests will require the use of areawide preventive measures. Appropriate suppressive measures will have to be applied in a coordinated manner against populations at strategic times and places during the seasonal cycles. In view of their long flight range, the elimination of *Heliothis* populations from regions will not be biologically or technically feasible. The goal would be to routinely maintain total populations in the areas of permanent survival below the level of significant damage. Suppression measures in the United States alone are unlikely to be adequate. In all probability, highly successful programs would require cooperation with Mexico. Populations may have to be managed in critical areas in Mexico early each season so as to minimize the number of immigrating moths. If populations in the overwintering areas could be rigidly regulated, the *Heliothis* problem would automatically be solved throughout the northern region, where immigrants from the overwintering areas, particularly *H. zea*, cause extensive losses each year.

Autocidal and parasite augmentation techniques are considered particularly promising for the areawide management of *Heliothis* populations. The technique to be evaluated in this report would involve the rearing and release of parasites that are specific or, at least, have a strong preference for the *Heliothis* species. The autocidal technique would involve the rearing and release of moths that have been altered genetically to prevent successful reproduction by their counterparts in natural populations. The possibility of the autocidal approach has been considered previously (Knipling 1982) and will not be analyzed in detail in this report. However, integrating the releases of parasites and genetically altered moths would seem to be highly advantageous for managing *Heliothis* populations. The possibility of making use of sex pheromones for continuous management of greatly reduced *Heliothis* populations should not be ruled out. But this possibility will not be evaluated in this report. While individual techniques may have advantages or disadvantages when employed alone, they offer unique opportunities for effective management when used concurrently or sequentially (see "Discussion," p. 204). The integration of other suppressive measures, such as controlling wild, early-season host plants in critical areas or applying biological insecticides, may also be useful (Stadelbacher 1982, Knipling and Stadelbacher 1983).

As a working hypothesis, *Heliothis* populations that have survived winter in the overwintering areas are assumed to be the most vulnerable to suppressive measures. Indeed, because of the great increase in size of *Heliothis* populations, large expansion in the cultivated-host-plant environment, and wide dispersal of the moths during the growing season, management of *Heliothis* populations by currently envisioned techniques and strategies may not be practical unless focused on the overwintered populations.

In the sections to follow, an appraisal will be made of the potential for regulating *Heliothis* populations by the timely release of one or more species of mass-reared parasites throughout the *Heliothis* overwintering areas.

Natural Biological Control Agents

Several hundred species of biological agents are known to prey on or parasitize various life stages of *Heliothis* (Whitcomb and Bell 1964, Vanden Bosch and Hagen 1966, Goodenough 1986). Effective management of *Heliothis* populations would seem to be virtually impossible were it not for the high mortality caused by these biological agents. The fact remains, however, that self-perpetuating populations of biological agents are often incapable of maintaining *Heliothis* species below serious damage levels. Thus, if biological organisms are to provide dependable control, it is my view that their numbers must be increased substantially by artificial means. As discussed elsewhere, factors inherent in nature's balancing scheme place limitations on the number of any single agent that can coexist with its hosts or prey and, hence, the role it can play in regulating the pest population. It is my opinion that the limitation in effectiveness of natural parasite populations is primarily due to their inadequate numbers at the right place and time during the host cycles. Therefore, much of the study presented in this chapter was focused on estimating the actual numbers of certain species of parasites that coexist with *Heliothis* populations and on quantitatively estimating the efficiencies of the parasites in inhibiting the growth of *Heliothis* populations in the natural environment. Based on the conclusion reached from those appraisals, estimates are made for the degree of control that may be expected if the parasite numbers are increased by artificial means early in the *Heliothis* season.

In the absence of definitive data on absolute numbers of the parasites and *Heliothis* hosts that coexist in natural environments, primarily deductive procedures were used in appraising the potential of the augmentation technique to regulate *Heliothis* populations. Two species of solitary larval parasites are included in this analysis: *Microplitis croceipes*, a hymenopterous species, and *Archytas marmoratus*, a tachinid species. The efficiency of *Eucelatoria bryani*, which exhibits multiparasitism behavior and may therefore mature several progeny in one host larva, was also evaluated. The estimated efficiencies of certain other *Heliothis* parasites have been considered in prior theoretical studies (Knipling 1971, 1977, 1979).

Basic *Heliothis* Population Model

To estimate absolute numbers of a given parasite that coexists with its host population by the procedures used in this study, it is necessary to first develop a host population model that is representative of the number and normal seasonal trends of the host population. According to Lewis (1970) and Hopper and King (1984), *M. croceipes* has a preference for second- to fourth-instar larvae. Therefore, key parameters in the model are the average number of these stages that each generation of *Heliothis* females produces during the seasonal cycles, the average number of host larvae that will be parasitized by coexisting *M. croceipes* females during their lifetime, and the survival rates to the adult stage for the immature parasites in the parasitized hosts and for the *Heliothis* larvae that escape parasitism by *M. croceipes*. Obviously, as discussed earlier in this report, a number of factors, biotic and abiotic, have positive or negative effects on the reproductive success of hosts and coexisting parasites. Together, these factors have a much greater regulatory effect on a host population than any single biological agent. In general, for the model proposed, the influence of all other factors is combined and no attempt is made to evaluate the importance of the various specific factors.

A simple way to portray the dynamics of an insect population during the season is to estimate the reproductive capability of a single overwintered female. Table 22 is proposed as representing the

Table 22
Basic model depicting the estimated values for key parameters in the population dynamics of *Heliothis* during a typical season (host plants do not include corn during the ear stage)

Parameter (per unit area)	Generation			
	1	2	3	4
Female *Heliothis*	1	4	13.5	35
2d– to 4th–instar larvae produced per female[1]	100	95	85	70
Total larvae	100	380	1,148	2,450
Survival of larvae to adults, percent[1]	8	7	6	4
Adult progeny	8	27	69	98
Increase rate	4.0	3.4	2.5	1.4

[1]Data are arbitrarily adjusted to reflect the increasing host density.

dynamics of *Heliothis* populations during normal seasons. Corn is excluded from the host plants in this model. Since *M. croceipes* parasitizes both *H. zea* and *H. virescens* larvae, no distinction between the dynamics of the two host species is attempted.

While the reproductive success of *Heliothis* females varies greatly from generation to generation, from habitat to habitat, and from ecosystem to ecosystem, I consider the estimates made for the different parameters to be reasonably realistic. Based on the fecundity of *Heliothis* species, we know that the survival rates in the various life stages are very low. Otherwise, populations would grow to extremely high levels. *Heliothis* females have the potential of depositing 1,000 eggs or more under laboratory conditions (Fye and McAda 1972). If only 1 percent normally survive to the adult stage, 10 adult progeny would be produced per female and would represent a 5-fold rate of increase. We know, however, that during most generations the increase rate is not 5-fold. If this rate of reproduction were to occur for four successive cycles, the potential would be for the population to have increased 625-fold by season's end. Thus, egg and larval densities on various crops would greatly exceed the numbers commonly observed. A maximum, seasonal production averaging nearly 100 adult progeny from each overwintered pair of moths is probably more realistic. This would mean a seasonal increase rate of 50-fold. Therefore, an average of five adult progeny per female per cycle, or an increase rate of 2.5-fold per cycle, is probably more representative of the average lifetime reproductive success of *Heliothis* females. The hypothetical population in table 22 increases about 50-fold over four generations. This growth rate per season is probably above average. Data on the survival rate of overwintering *Heliothis* pupae provide another useful clue for developing a typical *Heliothis* population model. While variable, the data presented by Johnson et al. (1986) indicate that the normal survival rate ranges from 2 to 3 percent. Thus, to maintain a steady density from year to year, the population must increase about 33- to 50-fold during the season. Such seasonal rates of increase would support the estimate that the average increase rate per cycle is about 2.5-fold.

It is a matter of judgment to estimate the average number of host larvae produced per female and the rates at which the different generations of larvae survive to the adult stage. These are very important estimates, however, and a great deal of thought was given to them. If we assume a 4-fold rate of increase as typical of the reproductive success during generation 1, we might consider the probable average number of host larvae that female *Heliothis* would have to produce and what the survival rates would have to be. If it is assumed, for example, that during generation 1, the number of second- to fourth-instar larvae produced per female normally averages 125, the survival rate would have to be as low as 6.4 percent to account for 8 adult progeny, or a 4-fold rate of increase. On the other hand, if we assume that an average of 75 such larvae are more likely to be produced per female, a survival rate of 10.7 percent would result in 8 adult progeny per female parent. The hypothetical model is based on an estimate of 100 host larvae per female during generation 1, and 8 percent survival of the larvae to the adult stage. I consider these values to be more realistic than the other two sets of values, but the key estimate is the assumed 4-fold rate of increase.

As *Heliothis* populations grow, the total effect of density-dependent suppression forces can be expected to increase for all of the life stages—eggs, pupae, larvae, and adults. Table 22 shows a diminishing increase rate to account for an increase in the effect of density-dependent suppression factors. However, other factors, such as weather conditions and the kind and amount of host plants, may be more important than the influence of host density on the rate of increase during any given cycle. For example, *H. zea* populations are likely to increase at a faster rate when reproduction occurs in corn ears than in other parts of the corn plant, as will be discussed later. But in general, I feel that the basic model will serve its purpose in enabling rough estimates to be made (1) of the size and growth trends of a coexisting natural *M. croceipes* population and (2) of the effect that an augmented *M. croceipes* population will have on the dynamics of the depicted *Heliothis* population.

Natural Coexistence Pattern of *Heliothis* and *Microplitis croceipes* Populations

Microplitis croceipes is a prevalent parasite of *Heliothis* larvae (Lewis and Brazzel 1968, Mueller and Phillips 1983, Stadelbacher et al. 1984, Puterka et al. 1985). The life history of the parasite has been investigated by Lewis (1970a), Danks et al. (1979), and others.

The first major objective was to estimate the number of larvae the females of a coexisting *M. croceipes* population will normally be capable of parasitizing during their lifetime in a natural environment. If the average parasitization capability of a single female is known within a reasonable range of error, it will be possible to extrapolate the number and proportion of a known larval host population that will be parasitized by any specified number of female parasites.

The parasite-host population model shown in table 23 indicates how efficient *M. croceipes* females must be to coexist successfully with a typical *Heliothis* population. The values assigned to the various parameters are largely self-explanatory, but some of the more important basic assumptions will be discussed briefly. The average number of larval hosts produced by *Heliothis* females and the survival rate to the adult stage are essentially the same as the values estimated for a single female and its progeny, as shown in table 22. In line with prior discussions, the lifetime host-finding capability of *M. croceipes* is assumed to be influenced little, if at all, by the host density so long as the density falls within the normal range for the species. Also, the rate at which the immature parasites survive to the adult stage is closely correlated with the rate at which *Heliothis* larvae that escape parasitism survive to the adult stage. Since the immature stages of *M. croceipes* complete development more rapidly than the host stages parasitized (Lewis 1970b), the rationale for this latter assumption might be questioned. But the parasitized host larvae may also be more vulnerable to predation and other hazards than are unparasitized larvae. Such a vulnerability could offset the advantages of more rapid parasite development. In any event, if the immature parasites have a lower or higher survival rate than do the unparasitized hosts, the parasite population would have to compensate by parasitizing, respectively, more or fewer hosts. The two organisms could not maintain a numerical balance for long if one or the other consistently produced significantly more adult progeny per female parent than the other. As discussed earlier in this publication, it is considered a biological truism that over the long term, the reproductive success of a solitary

Table 23
Basic model depicting the estimated coexistence pattern of *Microplitis croceipes* and *Heliothis* populations in a natural environment[1]

Parameter (1 acre)	**Generation**				
	1	**2**	**3**	**4**	**5**
Female *Heliothis*	25	94	300	734	1,051
Host larvae per female (2d–4th instar)	100	95	85	70	55
Total host larvae	2,500	8,930	25,500	51,380	57,805
Female M. croceipes	8.33	32	102	248	362
Ratio, adult parasites to adult hosts	1:3	1:3	1:3	1:3	1:3
Host larvae parasitized per female	75	71	63	53	41
Total host larvae parasitized	625	2,272	6,425	13,144	14,842
Parasitism, percent	25	25	25	25	25
Host larvae not parasitized	1,875	6,658	19,074	38,236	42,963
Survival rates to adults, percent	10	9	7.7	5.5	4
Adult *Heliothis*, next generation	188	600	1,468	2,103	1,718
Adult *M. croceipes*, next generation	63	204	495	723	594
Increase rate of host and parasite populations	3.8	3.2	2.4	1.4	.82

[1]The influence of the size and nature of an expanding host-plant environment is not considered in this model. The model is intended to reflect the total populations and population trends of the coexisting organisms in a closed ecosystem.

host-dependent parasite species will equal the reproductive success of its host.

The rate of parasitism due to *M. croceipes* is highly variable, judging from the reports of investigators previously cited. For the model, however, it is assumed to average 25 percent in an environment not subjected to the application of broad-spectrum insecticides. But any other average rate could be assumed without significantly influencing the estimated average host-finding capability the females must have. The rate of parasitism in a given ecosystem is a reflection of the relative numbers of coexisting adult parasites and hosts, and these numbers are determined by overall environmental influences. If parasitism averages 25 percent, the ratio of adult parasites to adult hosts should be about 1:3. This estimate is based on the hypothesis that the rates at which immature parasites and unparasitized host larvae survive to the respective adult stages tend to be equal.

According to the model, the *Heliothis* and *M. croceipes* populations would grow at about the same rate. The rate of increase in host and parasite populations is assumed to diminish over successive generations, but the rate of parasitism tends to remain constant. The average number of hosts parasitized per female parasite decreases as the host density increases, indicating a negative correlation between host density and the average host-finding capability of the female parasites. The total number of hosts parasitized in generation 5 exceeds the number parasitized during generation 1 by almost 25-fold.

Many models were developed and discarded before the model presented was selected. A special effort was made to develop a model of a *Heliothis* population that increases over successive cycles in accordance with the stipulation that the following two conditions apply concomitantly: (1) individual parasites will find and parasitize an increasing number, or even a constant number, of host larvae and (2) the *Heliothis* population will produce a decreasing number of larvae per female host because of increasing density-dependent suppressive forces. This effort led to the conclusion that such a model cannot be developed if realistic relative values are used for all key parameters. Highly distorted numerical relationships and obviously excessive rates of parasitism would occur before season's end.

Population biologists interested in the numerical relationship between *M. croceipes* and its *Heliothis* host might attempt to develop models depicting coexisting populations by drawing on their concepts of the dynamics and behavior of the two organisms. No doubt they would find the attempt interesting and revealing. Regardless of the simplicity or complexity of such models, it will soon become evident that the average number of hosts produced per *Heliothis* female, the average number of hosts parasitized per *M. croceipes* female, and the relative survival rates are highly sensitive parameters to be estimated. Models based on significant deviations (for example, 25 percent) in relative values for these key parameters will, within three to four cycles, result in numerical relationships, rates of parasitism, and relative growth rates for the two organisms that obviously would be inconsistent with reality.

As discussed previously, there is increasing evidence that for many parasite-host associations, the proportion of a host population that is parasitized does not increase as the host density increases. The model presented is in agreement with this evidence.

If the populations of both organisms normally decrease (or increase) at similar rates and the average rate of parasitism tends to remain constant or within rather narrow limits at normal host densities, a logical conclusion would be that the highly developed host-habitat-detection mechanism of the female parasites will enable them to find and parasitize nearly their normal quota of hosts at a wide range of host densities. This capability not only helps explain how parasitism works in natural populations but also has very important practical implications. It means that the most strategic time to gain maximum efficiency from a given number of released parasites is when *Heliothis* populations are at or near the lowest level. Normally, they are lowest in the spring, when they have been greatly reduced by winter mortality. Thus, the efficiency of the parasite augmentation technique, like that of the autocidal and sex-pheromone-based techniques, is pest-density dependent.

Influence of *Microplitis croceipes* Releases in *Heliothis* Ecosystems

If realistic estimates are made both of the average number of host larvae that *Heliothis* females produce during their lifetime and of the average number that are parasitized by *M. croceipes* females in a typical ecosystem containing coexisting populations of these two organisms, realistic estimates can also be made of the proportion of the larval population that is likely to be parasitized when a given number of parasites is released in the ecosystem. If the *Heliothis* population is to be managed by the release of parasites, much of the cost will depend on the costs of rearing and releasing the required number of parasites.

As shown in table 23, if parasitism averages 25 percent in natural populations, the ratio of adult *M. croceipes* to adult *Heliothis* is likely to be near 1:3. As previously noted, the female *Heliothis* is estimated to produce an average of 100 second- to fourth-instar larvae during the first cycle of the season. The female *M. croceipes* is estimated to parasitize an average of 75 such host larvae. These basic values will be used to estimate the rate of parasitism and its influence on the dynamics of a *Heliothis* population during the first generation if the ratio of adult *M. croceipes* to adult *Heliothis* is artificially increased to 3:1. This high ratio, nine times that estimated to result in 25 percent parasitism, can be expected to bring into play a number of factors affecting the population dynamics of *Heliothis* over successive generations.

For the purpose of this analysis, I assume that a closed *Heliothis* ecosystem has an overwintered population of 1,000 female *H. zea* and *H. virescens* per unit area. Most of the reproduction during generation 1 is assumed to occur on wild host plants. Subsequent generations are assumed to reproduce largely on cultivated crops, including cotton, soybeans, and tobacco. Corn is excluded in this analysis. Later, we will consider the possibility of controlling *H. zea* populations by releasing the parasite *Archytas marmoratus* in areas where corn is an important host crop. While the suppression model developed (table 24) is based on 1,000 male and 1,000 female moths per unit area, 100 million will be considered as a reasonable estimate of the number of adult *Heliothis* present during the first cycle in a typical southern State. This estimate is considerably higher than prior estimates made by Knipling (1971), Stadelbacher (1982), and Knipling and Stadelbacher (1983), and also two of 3 years' estimates made by Laster et al. (1987). But these last authors' estimate for one of the 3 years is, I believe, unrealistically high. I also believe that the estimates shown for generation 1 in table 24 are representative of the highest likely populations of overwintered *Heliothis* in the overwintering areas. In any event, based on the estimate of 100 million overwintered *Heliothis*, 300 million *M. croceipes* would have to be reared and then released to result in a 3:1 ratio during generation 1.

The model shown in table 24 presents the estimated influence of 3,000 released female *M. croceipes* on the rate of parasitism of host larvae produced by 1,000 female *Heliothis* and on the dynamics of the *Heliothis* population during generations 1 and 2. The estimates

Table 24
Suppression model depicting the estimated influence of *Microplitis croceipes* releases in *Heliothis* ecosystems (predominantly noncorn ecosystems)

Parameter (per unit area)	Generation		
	1	2	3[1]
Female *Heliothis*	1,000	656	225
Host larvae per female (2d–4th instar)[2]	100	105	—
Total host larvae	100,000	68,880	—
Female *M. croceipes* (releases in generation 1 only)	3,000	3,010	2,041
Ratio, adult parasites to adult hosts	3:1	4.6:1	9.1:1
Host larvae parasitized per female *M. croceipes*[3]	60	60	—
Total parasite-host encounters	180,000	180,600	—
Ratio, parasite-host encounters to host larvae present[4]	1.8:1	2.6:1	—
Parasitism, percent	83.6	92.7	—
Host larvae parasitized, number	83,600	63,778	—
Host larvae not parasitized, number	16,400	5,102	—
Survival, unparasitized larvae to adults, percent[2]	8	8.8	—
Survival, immature parasites to adults, percent[5]	7.2	6.4	—
Adult *Heliothis*, next generation (both sexes)	1,312	449	—
Adult *M. croceipes*, next generation (both sexes)	6,019	4,082	—

[1]In most *Heliothis* ecosystems, the volume of host-plant biomass will have expanded greatly by generation 3. Most of the data for generation 3 are therefore not estimated.
[2]Because of the declining host population, the average number of host larvae produced per female and the survival rate of unparasitized host larvae to the adult stage are assumed to increase slightly.
[3]The average number of host larvae parasitized per female is assumed to be 20 percent below normal because of the abnormally high parasite population.
[4]See table 1 to calculate the rate of parasitism by using the ratio of parasite-host encounters to hosts present.
[5]Superparasitism may cause premature mortality of some hosts. The survival rate of the immature parasite to the adult stage is assumed to be reduced by 10 percent during generation 1 and 20 percent during generation 2.

are based largely on the previously postulated parameter values; however, certain values were adjusted. The abnormally high *M. croceipes* population is expected to sustain a higher-than-normal rate of predation and, hence, a shorter-than-normal average life span. The average number of hosts parasitized per female *M. croceipes* is assumed to be reduced from 75 for a natural population to 60 for the augmented population. However, in view of the high parasite-to-host ratio, the total number of host larvae encountered and presumably parasitized versus the total present is estimated to be 1.8:1 during generation 1. Theoretically, this ratio means that 83.6 percent of the host larvae would be parasitized and that most of the parasitized larvae would be superparasitized. Superparasitization is arbitrarily assumed to reduce the survival rate of the immature parasite to the adult stage by 10 percent during generation 1 and 20 percent during generation 2. Nevertheless, the ratio of adult *M. croceipes* to adult *Heliothis* during generation 2 would increase to 4.6:1 and to a still higher ratio during generation 3. According to the values assigned for the average number of hosts produced by the female *Heliothis* and the average number parasitized by the female parasites, a 4.6:1 ratio would result in 92.7 percent parasitism during generation 2. Although a further increase in the parasite-to-host ratio would theoretically occur for generation 3, no estimate is made of the rate of parasitism because of the expanding plant-host environment.

Influence of the Expanding Plant Environment

Theoretically for some of the parasite-host associations included in this study, the rate of parasitism will increase progressively for several successive cycles following the release of enough parasites to raise the rate of parasitism beyond 50 percent. However, when *Heliothis* is the host, such an increase is unlikely because of the progressive increase in both the acreage and size of the host plants as the season advances. It seems desirable to analyze the probable significance of the increase in host-plant biomass on the rate at which *Heliothis* is parasitized.

While the rate of increase in host-plant biomass is likely to vary among *Heliothis* ecosystems and a thorough analysis would have to be made to determine the normal rate of increase, gross estimates, at least, can be made to assess the significance of this factor. My estimate is that by generation 3 of a *Heliothis* population, which typically will have increased by 12-fold (table 23), the host-plant biomass will have increased by 30-fold. Accordingly, under normal conditions for uncontrolled host populations, the parasite and host populations would be expected to increase somewhat proportionally to the increase in host-plant biomass over the first three host generations. Thus, the number of host larvae per unit of plant biomass would be expected to remain within a fairly narrow range, and the female parasites would be expected to find their normal quota of hosts. However, if parasites are released in the *Heliothis* ecosystem during generation 1 as previously proposed, my estimates are that the larval host population will decline steeply (table 24). By generation 3, this decline coupled with the estimated 30-fold increase in plant biomass would reduce the number of host larvae present per unit of plant material so drastically that, conceivably, the female parasites would have difficulty finding their normal quota of hosts.

I consider the plant-biomass factor to be so important that I hesitate to make an estimate of the rates of parasitism to expect during generations 3 and 4 in table 24. In effect, the *M. croceipes* females present during generation 3 would have to search about 36 times more plant material to find their normal quota of hosts. Despite their highly developed host-finding capabilities, I question whether under such conditions, the parasites would be capable of finding enough hosts to maintain parasitism at the level estimated for generation 2. Even so, however, the parasite-to-host ratio is estimated to be about 27 times higher than normal by generation 3, and the rate of parasitism should continue at a sufficiently high level to prevent the *Heliothis* population from increasing before season's end. If so, the overwintered population in year 2 should be low enough to be managed by the release of relatively few genetically altered moths during year 2 and subsequent years. An expanding plant-host environment should have no influence on the effectiveness of autocidal techniques. Therefore, the parasite augmentation and autocidal techniques used in combination should be much more effective and more efficient than either technique used alone.

Discussion

In prior general discussions, the view was expressed that certain fundamental principles govern the relationships that have evolved between virtually all hosts and host-dependent parasites. However, it was also noted that each parasite-host association is likely to have characteristics that are unique to the particular species involved. In our artificial agricultural environments, the parasites of *Heliothis* must cope with a greatly expanding host-plant environment as the season advances, because of the increase in acreage of cultivated host crops as well as the normal increase in size of all host plants. While natural parasite populations have undoubtedly adapted to this condition for uncontrolled *Heliothis* populations, it may have a highly significant negative influence on the role that parasite releases applied against total populations can play in regulating *Heliothis* populations.

There is little question, however, that *M. croceipes* is a highly efficient natural parasite. Its host specificity would seem to make it particularly valuable for use in augmentation programs. There is good reason to assume that the proposed released rate (table 24), which is estimated to raise the normal ratio of adult parasites to adult hosts by 9-fold during the first *Heliothis* generation, would have a major effect on the rates of parasitism and on the dynamics of *Heliothis* populations during the early generations. If so, the most important objective might be achieved: to reduce, on an areawide basis and by an ecologically acceptable method, natural *Heliothis* populations to a level that could be effectively regulated at minimal costs by the yearly release of genetically altered moths or of parasites.

Two estimates made in appraising the potential of *M. croceipes* to suppress *Heliothis* populations are of particular importance. The first is the estimate of the typical number of overwintered *Heliothis* adults in the southern States. The estimate of 100 million may be too high for most yearss. The other important estimate is the ratio of *M. croceipes* to adult *Heliothis* that would be required to result in high rates of parasitism and greatly diminished *Heliothis* populations by the end of the first year. I believe the chances are good that the proposed ratio of 3:1 would be adequate. Thus, in an area comparable to 1 southern State, 300 million *M. croceipes* would be released during generation 1, and this should have a major impact on the dynamics of *Heliothis* populations developing principally on wild host plants, especially in agricultural ecosystems where corn is not one of the major host crops. As will be considered later, the regulation of *H. zea* populations in an ecosystem where corn is a major host plant is likely to require special suppressive measures. The release of a better adapted parasite species and, perhaps, the release of genetically altered moths may be required to minimize successful reproduction of *H. zea* on corn.

Assuming that the release of 300 million *M. croceipes* will effectively suppress *Heliothis* populations in a typical southern State during the first year, we may consider the magnitude of the costs likely to be involved. Mass-rearing methods for *M. croceipes* have not been developed. Considerable progress has been made, however, on rearing *M. croceipes* for research purposes. According to personal communications with personnel of the U.S. Department of Agriculture, Southern Field Crop Insect Management Laboratory, Stoneville, MS, *M. croceipes* can be reared successfully on *Heliothis* larvae in the laboratory. Based on estimated numbers of parasites required to achieve high rates of parasitism, the development of mass-rearing methods for *M. croceipes* should be economically feasible, even if it must be reared on its normal host. If we assume that *M. croceipes* can eventually be mass-produced at a cost of $20 per 1,000 on *Heliothis* larvae, the cost of 300 million would be $6 million. Likely, however, the full benefits of a *Heliothis* management program would not be realized until after the first year. If a normal population of 100 million moths could be reduced to 10 million at a cost of $10 million during year 1, two options might be considered for further suppression and management in subsequent years. One option is to continue the release of *M. croceipes* against the reduced overwintered population. However, the efficacy of *M. croceipes* against abnormally low *Heliothis* populations in an expanding host-plant environment may be so low that the technique may not be highly efficient. The second option is to release genetically altered moths. This may prove to be a more effective and practical means of maintaining *Heliothis* populations at noneconomic levels on a year-to-year basis.

The release of the *H. virescens-H. subflexa* hybrid, developed by Laster (1972), may be the preferred method of regulating *H. virescens* populations on a continuing basis. When males of this hybrid mate with *H. virescens* females, the matings are sterile. When the females mate with normal *H. virescens* males, reproduction is normal but the male progeny are sterile and the female progeny are fertile. For regulating *H. zea* populations, the release of moths treated for optimum inherited sterility effects might be made. Using data obtained by Amoako-Atta et al. (1978) and Carpenter et al. (1983), as well as data from other investigators, on the effect of low doses of irradiation on reproduction by lepidoptera, Knipling (1982) estimated that a 9:1 ratio of *H. zea* adults treated for inherited sterility to normal moths would have a suppressive effect equal to five times the number of males that are 100 percent sterile. The inherited sterility technique may also be applicable for regulating *H. virescens* populations. If normal, overwintered *Heliothis* populations could be reduced to an average of about 10 million moths per southern State by the use of parasites during year 1, and if the release rate of genetically altered moths would have to be high enough to result in a ratio of 25:1, the cost for effective suppression would be very low. This ratio would require the release of 250 million *Heliothis*. In my opinion, *Heliothis* can eventually be mass-produced at a cost of $10 per 1,000. If it can, the cost for the insects would be only $2.5 million during year 2. But even if natural populations were merely held stable during year 2, losses due to *Heliothis* damage should be minimal, and natural control during the second winter should further reduce overwintered populations to a level that would not exceed 1 million moths. In such event, the routine release of as few as 50 million genetically altered moths per State each year would result in an overall 50:1 ratio of released moths to natural moths. If the effective ratio were only 10:1, the pest populations should be maintained below economically damaging levels, even though the intrinsic rate of increase of low-density *Heliothis* populations may be considerably above normal.

Thus, in theory, the prospects seem excellent that *Heliothis* populations could be effectively regulated on an areawide basis at less than 10 percent of the current losses under present control practices. However, in view of the migration factor, regionwide suppression programs may be required to ensure satisfactory results. Also, the release of more than one parasite species and supplemental releases of genetically altered moths may be necessary to effectively regulate *H. zea* populations in ecosystems where corn is a major host crop. In the section to follow, we will consider the role that *Archytas marmoratus* and *Eucelatoria bryani* might play when used for augmentation purposes and what the requirements might be to effectively manage *Heliothis* populations with these parasites.

Estimated Efficiency of *Archytas marmoratus* for Suppressing *Heliothis zea* Populations

Archytas marmoratus (Townsend) (fig. 7) is a tachinid species that develops in the larvae and pupae of noctuidae. The parasite is not *Heliothis* specific, which may be a disadvantage because some of the released parasites and their progeny are likely to be attracted to other hosts in the release areas. However, the parasite has a high affinity for *H. zea* larvae and has characteristics that would seem to make it an excellent candidate for augmentation purposes, especially in ecosystems where corn is an important host plant. Since the efficiency of parasites may depend on the kind of host plants involved, the release of several parasite species may be needed to effectively suppress *Heliothis* populations in various ecosystems.

I am grateful to H.E. Gross (Agricultural Research Service, Tifton, GA) for information, including unpublished data, on the biology and behavior of *A. marmoratus*. Hughes (1975) and Gross and Young (1984) obtained information on the biology and role of natural populations of the parasite. Gross and Johnson (1985) reported considerable progress on methods of rearing the parasite. There should be no technical barriers to the mass-production of the parasite in any numbers that might be required, and the costs should be reasonable.

As discussed earlier in this publication, the biology, behavior, and fecundity of parasites offer clues to their host-finding capability.

Figure 7. *Archytas marmoratus*. Like *Microplitis croceipes*, shown in figure 1, and several other parasites, *A. marmoratus* depends on *Heliothis* larvae for reproduction and survival. *A. marmoratus* deposits its larvae on plants infested with host larvae; and the parasite larvae, in turn, must find and parasitize the *Heliothis* larvae. An estimate of the efficiency of *A. marmoratus* indicates that the prospects are excellent that it could be economically mass-reared and released to achieve very high rates of parasitism of *H. zea* larvae developing on corn during the whorl and tassel stages of growth. *Heliothis* spp. attack a wide range of cultivated and wild host plants in diverse environments. Therefore, effectively managing *Heliothis* populations may require the mass-production and release of several different parasite species. The efficiencies of three parasite species are evaluated in this chapter.

Managing *Heliothis* populations on a regional or national scale, regardless of the control procedures employed, would require a program of unprecedented scope. This theoretical study suggests, however, that the release of appropriate parasite species at strategic times could provide a practical and ecologically safe means of suppressing populations of *H. zea* and *H. virescens* to low levels. By appropriately integrating parasite releases with autocidal techniques, it may be possible to manage popultions of these two *Heliothis* species at annual costs that would be less than one-tenth the cost of the losses they cause under present management guidelines. (Photograph, courtesy of Harry R. Gross, USDA Insect Biology and Management Systems Research, Tifton, GA.)

According to Hughes (1975), the larviparous female is capable of depositing from 1,800 to 2,800 larvae during its lifetime, depending on the temperature. The small larvae are deposited on host plants after a prelarviposition period of about 10 days. Although the larvae will parasitize earlier instars, fourth and fifth instars are more commonly parasitized (H.R. Gross, personal communication). Nettles and Burks (1975) reported that the frass of host larvae contains chemicals that stimulate larviposition by the female parasites. Gross (personal communication) believes that because they produce more frass than earlier instars, the larger larvae tend to attract most of the host-searching females.

The high fecundity of the parasite and its long prelarviposition period would suggest that a very low percentage of the immature progeny survive to the adult stage, even under the most favorable conditions for reproduction. Assuming that the female parasite is capable of depositing 2,300 larvae during its lifetime and that a 4-fold rate of increase, or 8 adults per female, is representative of very successful reproduction, the survival rate of the larvae to the adult stage would be only 0.35 percent. As discussed elsewhere in this report, the number of adult progeny produced per female parent tends to be essentially the same among solitary parasite species that share a dependency on the same host or host complex for reproduction, regardless of the host stages they parasitize. The more advanced the host stage attacked, the greater the probability of survival by the immature parasites and, hence, the fewer the number of hosts that must be parasitized. In the case of *M. croceipes*, the adult females have a preference for second- to fourth-instar larvae. It was estimated that under favorable conditions, when the *Heliothis* female produces an average of 100 such larvae, the female *M. croceipes* will parasitize an average of 75 larvae. Under similar conditions, the *Heliothis* female is estimated to produce an average of 50 fourth- and fifth-instar larvae and the *A. marmoratus* female to parasitize an average of 45 such larvae. But the average number of adult progeny produced per female parent is likely to be similar for both *M. croceipes* and *A. marmoratus.*

The model shown in table 25 is assumed to be representative of the coexistence pattern of *H. zea* and *A. marmoratus* populations developing in corn before the ear stage. The rate of parasitism in natural populations is assumed to average 10 percent. The survival rates for the unparasitized host larvae and the parasite larvae to the adult stage are assumed to be essentially equal. Therefore, the ratio of adult parasites to adult hosts when parasitism averages 10 percent is estimated to be about 1:9.

Table 25
Basic model depicting the estimated coexistence pattern of *Archytas marmoratus* and *Heliothis zea* populations in early-whorl to late-tassel corn

Parameter (1 acre)	Generation	
	1	2
Corn plants, number	25,000	25,000
Female *H. zea*	100	360
Larvae per female (4th–5th instar)[1]	50	45
Total host larvae	5,000	16,200
Corn plants infested, percent	20	65
Female *A. marmoratus*	11	40
Ratio, adult parasites to adult hosts	1:9	1:9
Host larvae parasitized per female[1]	45	40
Host larvae parasitized, number	500	1,600
Host larvae parasitized, percent	10	10
Host larvae not parasitized	4,500	14,600
Survival, unparasitized larvae to adults, percent[1]	16	14
Adult *H. zea*, next generation	720	2,044
Survival, parasite larvae to adult parasites, percent[1]	16	14
Adult *A. marmoratus*, next generation	80	224
Increase rate for host and parasite populations	3.6	2.8

[1]Because of the increase in host and parasite densities, the average number of host larvae produced per *H. zea* female, the survival rates of the parasite larvae and unparasitized host larvae, and the average number of hosts parasitized per female are assumed to decrease. The survival rates of the parasite larvae and unparasitized host larvae are assumed to be equal.

A survival rate of 16 percent for the parasite larvae and unparasitized host larvae to the respective adult stages would mean a 3.6-fold rate of increase for both organisms. The average number of host larvae produced per female *H. zea* and the average number parasitized per female parasite are assumed to decrease slightly during generation 2 because of the increases in host and parasite densities. However, the percentage of corn plants infested would increase from 20 percent to 65 percent. The model is probably representative of rather severe infestations of *H. zea* in early-season corn. The parasite would reproduce successfully but would contribute little to the control of *H. zea*. Such a limited effect is typical of many parasite-host associations.

As in the case of *M. croceipes*, the average number of hosts parasitized per female parasite is assumed not to increase with an increase in the host density. The total number of host larvae parasitized by the parasite population increases linearly with the increase in the host density, but the proportion of the total larval population parasitized tends to remain constant. If it were assumed that the number of hosts parasitized per female increases proportionally to the increase in the host density, the two populations would become grossly unbalanced by generation 3. By generation 2, about 32 percent of the host larvae would be parasitized. Comparable survival rates for the immature parasites and unparasitized host larvae to the respective adult stages would then cause a shift in the ratio of parasites to hosts from the assumed normal of 1:9 to about 1:2.1. A higher rate of parasitism and a further distortion in the numerical ratio could then be expected in generation 3. Thus, the models developed lead to the conclusion that as noted for many other parasite-host associations, a positive correlation between host density and host-finding capability of the female parasites is unlikely.

The coexistence model in table 25 is proposed as reasonably representative of the numerical relationship between *A. marmoratus* and its *Heliothis* host. As noted, other hosts are also parasitized by *A. marmoratus*, but I would expect the numerical relationship between *A. marmoratus* and other suitable hosts to be somewhat similar. The most important parameters are the estimated relative numbers of host larvae produced by the *Heliothis* females and host larvae parasitized by coexisting *A. marmoratus* females and the relative survival rates of the immature parasites and unparasitized host larvae to the respective adult stages. The estimated values for these key parameters will be used with some modification to calculate the effect of augmented *A. marmoratus* populations.

Influence of *Archytas marmoratus* Releases on *Heliothis zea* Populations During the Whorl and Tassel Stages of Corn

A suppression model showing the theoretical effects of *Archytas marmoratus* releases on a *Heliothis zea* population is shown in table 26. The *H. zea* population is assumed to average 100 females per acre during generation 1. Each female is assumed to produce an average of 50 large larvae, so the resulting larval density is 5,000 per acre. If the corn plant population is 25,000, the infestation rate would be 20 percent. *A. marmoratus* is assumed to be released at the rate of 300 females per acre. The ratio of adult parasites to adult hosts would thus be 3:1, or about 27 times the ratio estimated to result in 10 percent parasitism of natural larval populations early in the season. The abnormally high parasite population is assumed to result in above-normal predation and a shorter average life span. Therefore, the average number of host larvae parasitized per female is assumed to be reduced by 20 percent from a normal average of 45 larvae to 36. However, the high parasite-to-host ratio would mean that the total number of parasite-host encounters (10,800) would exceed the number of host larvae present by a ratio of 2.2:1. According to the probability data shown in table 1, this ratio would result in 89.5 percent parasitism, or

Table 26
Suppression model depicting the influence of *Archytas marmoratus* releases in *Heliothis zea* ecosystems containing corn plants

Parameter (1 acre)	Host generation	
	1	2
Corn plants	25,000	25,000
Female *H. zea*	100	42
Large host larvae per female (4th–5th instar)[1]	50	52
Total host larvae	5,000	2,184
Corn plants infested, percent	20	8.7
Female *A. marmoratus* released (generation 1)	300	268
Ratio, adult parasites to adult hosts	3:1	6.4:1
Larvae parasitized per female parasite	36	27
Total parasite-host encounters	10,800	7,236
Ratio parasite-host encounters to host larvae present[2]	2.2:1	3.3:1
Parasitism, percent	89.5	96.5
Host larvae parasitized, number	4,475	2,108
Host larvae not parasitized, number	525	76
Survival, unparasitized larvae to adult *Heliothis*, percent[3]	16	18
Adult *Heliothis*, next generation	84	14
Survival, parasite larvae to adult parasites, percent[4]	12	10
Adult *A. marmoratus*, next generation	537	211

[1]In view of the decline in host density, the average number of host larvae produced per female is assumed to increase slightly.
[2]See table 1 for calculating the rate of parasitism from the ratio of parasitie-host encounters to hosts present.
[3]In view of the decline in host density, the survival rate of the host larvae is assumed to increase.
[4]In view of the high degree of superparasitism expected, the survival rate of the immature parasites is assumed to decrease.

10.5 percent escape from parasitism. A high proportion of the larvae would be parasitized more than once. According to Hughes (1975) and Gross and Young (1984), excessive reparasitization results in above-normal mortality of the developing parasite larvae. How much value to assign to this negative factor is a matter of speculation. I arbitrarily assume that the survival rates of the immature parasites will be reduced by 25 percent during generation 1. Even so, the ratio of adult parasites to adult hosts would more than double during generation 2. For a *Heliothis zea* population that would normally be expected to increase 3.6-fold in generation 1, a parasitization rate near 90 percent would theoretically cause a decline of nearly 60 percent in the adult host population during generation 2, and a similar decline in the percentage of infested plants. The average number of host larvae parasitized per female *A. marmoratus* during generation 2 is assumed to decline to 27 because of the lower host density. This adjustment is probably very conservative. Nevertheless, in view of the high parasite-to-host ratio, the rate of parasitism would theoretically increase to 96.5 percent during generation 2.

Theoretical Results Versus Experimental Data

The theoretical models presented for the *H. zea* (corn earworm) and *A. marmoratus* association were developed before recent experimental results obtained by Gross (1988) were published. Gross investigated the feasibility of controlling corn earworm (CEW) larvae on corn in the whorl and tassel stages of growth with *A. marmoratus* (AM) placed on the corn plants. In brief, Gross conducted 3 experiments in which he placed from 20 to 24 AM larvae (average of 22 larvae) on corn plants harboring a single third-, fourth-, or fifth-instar CEW larva. The AM larvae had been extracted mechanically from gravid females. After 24 to 32 hours' exposure of the CEW larvae to the AM larvae, the rate of parasitism ranged from 56.5 percent to 79.8 percent and averaged 69 percent for the three experiments.

An opportunity is thus available to compare Gross' results with the theoretical results presented in table 26, provided a good estimate can be made of the number of larvae likely to be deposited on the infested plants by the released AM. As noted earlier, Hughes (1975) found that the female AM is capable of depositing 1,800 to 2,800 larvae, depending on the temperature. Since the parasites are assumed to be released when conditions for their reproduction are normal, the average fecundity is assumed to be 2,300 larvae per female. However, as previously mentioned, I estimate that due to natural mortality, the number of eggs or larvae actually deposited by female parasites in natural environments will usually be near 33.3 percent of the fecundity. Also, due to the abnormally high parasite density in augmentation programs, I assume that the average number of larvae deposited by the released females will be 20 percent below normal. Therefore, the released AM females are each estimated to deposit an average of 613 larvae.

On this basis, 300 AM females would deposit 183,900 larvae on 5,000 infested corn plants during generation 1. So an average of approximately 37 larvae would be deposited per infested plant. Gross' data indicate that 22 larvae per plant will result in about 69 percent parasitism. Therefore, 37 per plant would be expected to result in about $69 + (15/22)(0.69(100 - 69)) = 69 + 14.6 = 83.6$ percent parasitism. This percentage compares reasonably well with the 89.5 percent parasitism estimated for generation 1 in the suppression model. It should be noted, however, that the CEW larvae in experiments conducted by Gross were exposed to the parasites for only about 24 to 32 hours. The estimated rate of parasitism depicted in the model is based on the premise that the CEW larvae will be exposed to parasitism until they grow to maturity. Thus, the CEW larvae in natural populations would be subject to parasitism for at least 2 to 3 days, or about two times as long as the exposure permitted by Gross. It seems likely, therefore, that the estimate of 89.5 percent parasitism due to the released AM is a realistic, if not a very conservative, estimate.

The theoretical results projected for generation 2 also can be compared with the experimental data. According to the model, 268 AM females would be present when the plants infested with CEW larvae number 2,184. Again, assuming that each female will deposit an average of 613 larvae, the total number of AM larvae would be 164,284. Thus, the density of AM larvae would average 75 per infested plant, or about two times the density estimated to result in 89.5 percent parasitism. The theoretical model projects a parasitization rate of 96.5 percent. Considering the high parasite larval density and the longer exposure time of the host larvae to the parasite larvae under field conditions, the actual rate of parasitism may be higher than the theoretical.

The experimental data obtained by Gross suggest that the estimated rates of parasitism presented in table 26 are not likely to be too high. Since the theoretical data tend to agree with Gross' experimental data, there is reason to have considerable confidence in the general procedure used to estimate the efficiency *A. marmoratus* and of other parasites included in this investigation.

Further Discussion of Theoretical Results

Like all hypothetical models, those presented in this publication contain many estimated and assumed values that cannot be readily verified, because of insufficient field data. Therefore, we can only view such models from a broad perspective to consider whether they seem reasonable. There is little question that 300 female *A. marmoratus* per acre during the first generation would represent a very high population compared with normal numbers in natural populations. Therefore, the rate of parasitism calculated for generation 1 would seem plausible. I feel that realistic adjustments have been made for the various parameters during generation 2 and that the rate of parasitism estimated for generation 2 is also realistic. As discussed, the data obtained by Gross tend to support the theoretical results.

The release of *A. marmoratus* in the numbers proposed should have a major impact on *H. zea* populations in a predominantly corn ecosystem. An uncontrolled *H. zea* population, as shown in table 25, is assumed to increase from 100 adult females per acre in generation 1 to approximately 1,000 per acre by generation 3. The suppressed population, according to the model (table 26), would be reduced to an average of seven females per acre by generation 3. Therefore, the population in relation to an uncontrolled population would be reduced by more than 99 percent. It is recognized, however, that the only way to determine whether the theoretical results are realistic is to conduct suitable experiments.

Even though *H. zea* populations may be suppressed to very low levels by generation 3, without some means of further control, they may increase to near normal numbers by season's end if they reproduce in ears of corn. There, the larvae are well protected from natural biological control agents. If the population numbered 10 females per acre and each deposited an average of 500 eggs, a total of 5,000 eggs would be deposited in 1 acre. Assuming 50 percent hatch, the ear infestation level may still be on the order of 10 percent. If the population increased by 10-fold, or from 10 to 100 females per acre, and migrated to corn in the ear stage, enough adult progeny may then be present to cause near-100-percent infestation in corn ears during generation 4.

No doubt, normal, uncontrolled populations generally produce an excess of eggs on ears of corn. Egg densities may commonly exceed an average of 10 per ear on corn silks, although cannibalism among the larvae is likely to allow only 1 larva per ear to complete development. Even if cannabalism did not occur and about 90 percent of the eggs were destroyed by biological agents or were unable to hatch because of sterility, one larva per ear may still be able to complete development. Thus, effective *H. zea* control on corn is likely to be very demanding by any biological procedures we now could apply. It may be necessary to reduce adult populations to such low levels that the number of eggs they deposit will be considerably lower than the number of corn ears. If this can be accomplished and if most of the eggs that are deposited cannot hatch, a high degree of control can be expected. A likely strategy, therefore, is to release the parasites to reduce the generation-3 population to a low level, and then release moths treated for optimum inherited sterility effects to limit reproduction in ears of corn. This strategy, which will be discussed in greater detail in the next section, would seem to be particularly effective in minimizing the number of *H. zea* progeny that could develop in corn ears and mature before season's end.

For this theoretical analysis of the feasibility of integrating parasite and sterile moth releases, an *H. zea* population is assumed to have been reduced by released parasites to an average of 10 males and 10 females per acre by the time corn has reached the ear stage. This reduced population will be designated as generation 3. The objective would then be to release enough *H. zea* males treated for inherited sterility effects to suppress reproduction during generations 3 and 4. We will assume the release of 100 treated males per acre during generation 3. The theoretical effects will be described briefly. As a result of the release, the ratio of treated males to normal males will be 10:1. If the males are fully competitive, about 9 (actually 9.1) of the females will mate with released males and 1 will mate with a wild normal male. Multiple mating occurs in *Heliothis* but is not likely to change the results significantly. Mating between normal females and normal males is assumed to result in a 10-fold increase in adult progeny. However, when normal females mate with treated males, 50 percent of the eggs they deposit will not hatch. The larvae that do hatch are assumed to develop normally; however, for all practical purposes, the adult progeny are sterile. Therefore, the population per acre during generation 4 would consist of 10 normal males, 10 normal females, 45 sterile males, and 45 sterile females. Thus, the sterile-to-fertile ratio would be 4.5:1. At this ratio, 1.8 normal females would be expected to mate with normal males ($1/5.5 \times 10 = 1.8$). The remaining normal females would mate with sterile males. From the normal matings, we would expect a potential adult population of 18 normal males and 18 normal females per acre. However, these are assumed to be the potential overwintering population, and perhaps no more than 5 percent will survive the winter. The overwintered moth population during year 2 may therefore average about two per acre. In contrast, if the hypothetical population during generation 3 were not subjected to the released males, the population during generation 4 would average 100 normal males and 100 normal females per acre. The potential overwintering population would average 1,000 normal males and 1,000 normal females. Thus, the released treated males would, in theory, provide about 99 percent control.

The theoretical influence of the release of males treated for optimum inherited sterility effects on reproduction by emigrating moths would be of special significance. It is assumed that a regional suppression program is involved and that the suppression measures are applied in the overwintering areas. However, there is likely to be considerable long-distance movement of moths to non-overwintering areas during generation 3. Even so, if matings occur before emigration and if the progeny from the normal matings and from the matings between the irradiated males and normal females disperse in the same manner, the influence on the moths in the areas of spread should be essentially as described for the population that does not disperse.

The scenario described suggests that the release of irradiated moths after the release of parasites would reduce *H. zea* reproduction to very low levels. In such event, the natural mortality during the winter should lead to a near-extinct overwintered population at the start of year 2. The release of relatively few genetically altered moths during year 2 might then be the most practical means of maintaining very low natural populations on a continuing basis. While actual results are not likely to equal the theoretical, the total overwintered population should be so reduced within 2 or 3 years that the release of perhaps as few as 10 million adults per average southern State would ensure that the populations will remain below damaging levels. Such a program may cost less than 10 percent of the cost of losses the pest causes under current management practices in the overwintering areas and should also prevent all losses in the non-overwintering areas. The release of parasites against low *Heliothis zea* population may, however, be equally or more effective.

There is little question that *H. zea* is a particularly formidable pest and that extraordinary suppressive measures would be necessary to cope with it in an effective and ecologically acceptable manner in our abnormal agricultural environment. But there is also little question in my mind, that scientists could develop such measures. However, the measures would have to be applied in a fully coordinated manner and against total populations at strategic times and places. Since the feasibility of employing the techniques proposed will depend on the number of *H. zea* in natural populations, some idea of the number of overwintering *Heliothis* in natural populations in the United States is a matter of major importance. Estimates of the magnitude of overwintered populations will be the subject of discussion later in this chapter.

Influence of *Eucelatoria bryani* Releases on *Heliothis* Populations

Eucelatoria bryani (Sabrosky) is a primary larval parasite of *Heliothis zea* and *H. virescens*. It is a larviparous tachinid that parasitizes fourth to fifth instars (Bryan et al. 1970). The female parasite deposits more than one larva per host, thus exhibiting "multiparasitism" behavior. *Eucelatoria bryani* (Eb) is prevalent in the southwestern region of the United States.

Data obtained by Puterka et al. (1985) indicate that the average rate of parasitism probably does not exceed 5 percent. However, the ratio of adult parasites to adult hosts would be higher than the rate of larval parasitism would indicate because generally, more than one parasite can develop to maturity at the expense of one parasitized host. The rate of parasitism caused by a parasite species may be low in nature, but this is not necessarily indicative of the parasite's potential to be successfully used for augmentation purposes. It should be noted that rates of parasitism that are based on field-collected host larvae in various stages of development would not be accurate for parasites that parasitize only late-stage larvae. Thus, the data available for *E. bryani* may not reflect its true effectiveness.

Important basic research has been conducted by Nettles et al. (1980) on methods of rearing Eb. It can be reared on artificial diets after the first-instar larvae have developed in host larvae for 24–32 hours. Various aspects of the biology of Eb have been investigated by Jackson et al. (1969) and Bryan et al. (1970, 1972).

There is reason to believe (W.C. Nettles, personal communication) that Eb can eventually be mass-produced at very low cost. Therefore, it might prove to be the most promising *Heliothis* parasite for augmentation purposes in environments where it performs effectively during the regular growing season. Survival of a parasite species from one season to the next would be advantageous but is not essential for successful suppression of host populations by the augmentation technique if the adults perform normally during the growing season.

Estimated Host-Parasitization Capability of *Eucelatoria bryani*

A theoretical appraisal of the efficiency of *E. bryani* to parasitize *Heliothis* was reported earlier (Knipling 1977), and it was based on the estimated number of adults per acre required to parasitize 50 percent of *Heliothis* larvae. A reappraisal based on the procedures adopted for this investigation is presented in this section.

A model to indicate the normal coexistence pattern of Eb and *Heliothis* populations will not be presented. I estimate, however, that under normal, favorable conditions, the Eb female will parasitize an average of about 12 fifth- to sixth-instar *Heliothis* larvae during its lifetime. The rationale for this estimate can be understood by comparing the efficiency of Eb with that of *M. croceipes* (see table 23). The *M. croceipes* female, which deposits 1 egg per host, is estimated to parasitize an average of 75 second- to fourth-instar larvae under favorable conditions. Eb parasitizes older larvae, and each parasitized larva has the potential of yielding two to three adult parasite progeny (W.C. Nettles, personal communication). For the reappraisal, I assume an average of 2.5 adult progeny per parasitized host. Generally, the larvae of *Heliothis* spp. are highly vulnerable to predation. I estimate that under normal conditions on most host plants, only about 40 percent of second- to fourth-instar larvae survive to become fifth- to sixth-instar larvae. An average of 12 such large larvae parasitized by Eb would thus represent a reproduction success comparable to that of *M. croceipes*. The rationale for these estimates is consistent with the theory that in terms of the average number of adults produced per female parent, two or more parasite species that are totally dependent on the same host species for reproduction will be comparably successful under similar environmental conditions, regardless of the rates at which they parasitize natural host populations or of differences they may have in parasitization behavior.

Suppression Model

The model shown in table 27 is proposed as representative of the influence of Eb releases on the rates of parasitism and on the dynamics of *Heliothis* populations. It is based on the assumption that enough Eb females will be released during generation 1 to achieve a 7.5:1 ratio of adult parasites to adult hosts. A 3:1 ratio of adult *M. croceipes* to adult *Heliothis* is estimated to result in a similar rate of parasitism. As will be discussed later, however, the relative costs of rearing the two parasite species may favor Eb by a wide margin.

In natural populations, the Eb female is estimated to parasitize an average of 12 large host larvae during its lifetime and, therefore, to have the potential of producing an average of 30 adult progeny. However, natural mortality will greatly reduce the average. Also, for the augmented population predation is expected to be above normal, so the Eb female is estimated to parasitize an average of 10 large host larvae during generation 1. The average is further reduced to 9 per female during generation 2 because of the further increase in the parasite population. In contrast, the average

Table 27
Suppression model depicting the estimated influence of *Eucelatoria bryani* (Eb) releases in *Heliothis* ecosystems (except in areas or periods when corn ears are developing)

Parameter (per unit area)	Generation[1]		
	1	2	[2]3
Female *Heliothis*	1,000	750	200
Host larvae per female (5th–6th instar)[3]	40	41	45
Total host larvae	40,000	30,750	9,000
Eb females released (generation 1)	7,500	10,625	9,494
Ratio, adult Eb to adult *Heliothis*	7.5:1	14.2:1	47.5:1
Host larvae parasitized per Eb female[4]	10	9	—
Total parasite-host encounters	75,000	95,625	—
Ratio, parasite-host encounters to host larvae present[5]	1.9:1	3.1:1	—
Parasitism, percent	85	95	—
Host larvae parasitized and superparasitized	34,000	29,212	—
Host larvae not parasitized	6,000	1,538	—
Potential adult Eb progeny per parasitized larva[6]	2.5	2.5	—
Potential adult Eb progeny	85,000	73,030	—
Survival of unparasitized larvae to adults, percent[3]	25	26	—
Survival of Eb larvae to adults, percent[3]	25	26	—
Adult *Heliothis*, next generation	1,500	400	—
Adult Eb, next generation	21,250	18,988	—

[1]The increase in plant size and the probability that the acreage of cultivated host crops increases as the season advances are not considered in this model. The possible adverse effects of these factors on the average number of host larvae parasitized per female parasite might be partly offset by a higher parasite-to-host ratio than assumed.
[2]Dashes indicate that the data are not estimated.
[3]Arbitrary adjustments are made in the number of larvae produced per *Heliothis* female and in the survival rates because of the declining host density.
[4]The average number of hosts parasitized per Eb female during generation 1 is assumed to be about 20 percent below normal because of the high parasite density resulting from the releases. The lower average during generation 2 is assumed because of the increase in the parasite population.
[5]See table 1 to calculate the rates of parasitism and escape from parasitism by using the ratio of parasite-host encounters to hosts present.
[6]The higher the rate of parasitism, the greater will be the probability of superparasitism. If the superparasitized larvae have the potential to produce more parasite progeny than assumed, the ratio of adult parasites to adult hosts during generations 2 and 3 would be higher than estimated.

number of large host larvae produced per female *Heliothis* and the survival rates are assumed to increase because of the declining host population. While these adjustments would favor the host population, they are, nevertheless, inadequate to compensate for the large increases in the parasite-to-host ratio. Therefore, the rate of parasitism increases.

In general, the theoretical results shown in table 27 are similar to those in other suppression models developed to estimate the influence of augmented parasite populations on host populations. The key parameters are the average number of host larvae produced by the female *Heliothis*, the average number of host larvae parasitized by the coexisting female parasites, and the relative survival rates of the unparasitized host larvae and parasite larvae to the respective adult stages. The assumed average number of potential parasite progeny per parasitized larvae is, however, a parameter of special significance in the case of Eb. Most of the other parasite-host complexes considered in this investigation involve solitary larval parasites.

Discussion of Results

If the suppression model accurately depicts the influence of the Eb releases, an initial parasite-to-host ratio of 7.5:1 should lead to a strong suppression of the *Heliothis* population. The high rate of parasitism during generation 1 would theoretically result in a higher parasite-to-host ratio and a higher rate of parasitism during generation 2. Therefore, the release of a large number of parasites into a *Heliothis* ecosystem would greatly alter not only the normal numerical relationship between the parasite and host populations but also the spatial relationships in the changing plant-host environment in the subsequent generations. The influence of the expanding host-plant environment has already been discussed in relation to the influence of *M. croceipes* releases.

As noted in chapters 2 and 3, certain factors and fundamental principles govern the parasitization processes for all parasite-host associations. However, some associations have evolved rather unique biological and behavioral characteristics that must be critically analyzed if reliable estimates are to be made about the influence of parasite releases. In the case of Eb, the multiparasitism behavior is particularly important. In natural populations, the probability of reparasitization is so low that it is not likely to greatly influence the reproductive success of a parasite population. If natural parasitism averages 5 percent, only about 2.5 percent of the parasitized larvae would be reencountered by chance and, presumably, reparasitized. So relatively few additional adult progeny would be produced because of reparasitism. However, since the probability of reparasitism depends on the rate of parasitism that has already occurred, a substantial increase in the average number of adult parasites per parasitized larva could be expected when large numbers of Eb parasites are added to *Heliothis* ecosystems. According to Ziser et al. (1977) and Martin et al. (1989), larvae heavily reparasitized by Eb may each yield 10 or more adult parasites, although the size and fecundity of the females will be reduced. In contrast, larvae excessively reparasitized by a solitary parasite species may each yield less than one adult parasite because of intraspecific competition and parasite mortality. Therefore, a high rate of superparasitism is considered a strong negative factor when solitary species are used for augmentation purposes but could actually be a strong positive factor when Eb is used.

Since intraspecific competition seems not to be a mechanism for controlling the size of Eb populations, this parasite should have a considerable advantage over solitary parasite species in augmentation programs. It is puzzling to me, however, why the rate of natural parasitism due to Eb tends to remain so low. Some factor not readily apparent must control the size of Eb populations and tend to maintain the ratio of adult Eb to adult hosts in self-perpetuating populations at a low level. As all of the suppression models involving the release of solitary parasites indicate, the encounter of parasites with previously parasitized hosts and the intraspecific competition that results when such hosts are reparasitized will limit the rate of parasitism that occurs, especially in subsequent generations. But if larvae superparasitized by Eb produce more parasite progeny than those parasitized only once, *Heliothis* should be much more vulnerable to suppression by releases of Eb than by releases of *M. croceipes* or *A. marmoratus*.

In generation 1, according to the parameters shown in table 27, the Eb females have the potential to attack host larvae 75,000 times, or an average of 1.9 times per larva. If each attack resulted in 2.5 potential progeny, the potential adult Eb population during generation 2 would be 187,500 instead of 85,000, the population shown in the model. If 25 percent survived, the adult parasite population during generation 2 would number 46,875 instead of 21,250. In such case, the larvae during generation 2 would be parasitized 210,933 times, and the rate of parasitism would exceed 99 percent. However, the model is based on the conservative assumption that only 2.5 adult progeny will be produced per parasitized host whether it is parasitized once or more than once. The same assumption is made for each generation. Thus, the model may be based on very conservative estimates of the rates of parasitism that would result.

We may judge the reasonableness of the estimated rates of parasitism shown in what may be a very conservative model (table 27). If natural parasitism averages 5 percent, 2,000 of the estimated 40,000 large larvae would be parasitized. If 25 percent of the parasitized larvae survive and produce an average of 2.5 parasite progeny, the total number of adult parasites would be 1,250. Survival of 25 percent of the 38,000 larvae that escape parasitism would mean 9,500 adult *Heliothis*. Thus, an estimated parasite-to-host ratio of 1,250:9,500 or 1:7.6 can be expected to result in 5 percent parasitism. However, if enough Eb are released to raise the ratio to 7.5:1, or 57-fold, the estimate of 85 percent parasitism during generation 1 would seem plausible. The host larvae would be parasitized an average of about two times. But assuming that each parasitized host larva has the potential of yielding only 2.5 adult parasite progeny and assuming comparable survival rates for the immature parasite progeny and unparasitized host larvae, the ratio of adult parasites to adult *Heliothis* would increase from 7.5:1 during generation 1 to 14.2:1 during generation 2. The larval host population would not decline very much. Nevertheless, the estimate of 95 percent parasitism during generation 2 would seem reasonable, considering that a ratio of 14.2:1 is 108 times the ratio estimated to result in 5 percent parasitism and almost 2 times the ratio estimated to result in 85 percent parasitism. Considerable expansion of the volume of plant host material may occur in generation 2 and influence the average number of larvae parasitized per female. On the other hand, the potential average number of adult progeny per parasitized host larvae may be much higher than 2.5, because of superparasitism during generation 1. Therefore, the rate of parasitism and its influence on the *Heliothis* population may be even greater than depicted in the model.

In generation 3, however, major changes in the numerical and spatial relationships between the parasites, host insects, and host plants could be expected. For these reasons, an estimate of the rate of parasitism might have little validity. We can, however, indicate the magnitude of the changes in relative numbers of parasites and hosts and the amount of plant material that is likely to be present. If parasitism reaches a level of 95 percent during generation 2, a sharp decline in the *Heliothis* population could be expected. The adult and larval populations respectively, are estimated to be 80 and 77.5 percent below the populations present in generation 1. Based on the assumption that each parasitized host larva during generation 2 has the potential to yield 2.5 adult parasite progeny, the ratio of adult parasites to adult hosts would be 47.5:1 during generation 3, or about 360 times the ratio estimated to result in 5 percent parasitism. But since the host larvae during generation 2 would be parasitized an average of about three times, the average number of progeny per parasitized larva is likely to be considerably higher than 2.5. If it is higher, the parasite-to-host ratio would be higher than 47.5:1. On the negative side, if the larval host population is 77.5 percent below the normal low level and the volume of *Heliothis* plant material has expanded by 20-fold or possibly more, the female parasites would have to search about 100 times the volume of plant material during generation 3 than the volume during generation 1. Thus, although the ratio of adult parasites to adult hosts is assumed to have increased by about six times by generation 3, the average host-finding capability may not be high enough to compensate for the much lower number of larvae per volume of plant material.

Without question, as discussed in chapter 2, the ability of parasites to find their hosts by responding to kairomone signals produced by the hosts greatly minimizes low host density as a factor governing the host-finding efficiency of parasites. But it can also be rationalized that in the case of *Heliothis*, host populations per volume of plant material may reach such low levels by generation 3 that host finding becomes a critical survival factor for the parasites present. Therefore, no estimate is made of the average host-finding capability of the Eb females during generations 3 and 4. Theoretically, however, the ratio of parasites to hosts would be so high that if the females parasitized an average of as few as two or three host larvae, the rate of parasitism should still range from 90 to 95 percent during generations 3 and 4 and at least prevent an increase in the *Heliothis* population.

An uncontrolled *Heliothis* population under normal conditions is estimated to increase by 33-fold or more during the season. Then, if 3.3 percent normally survives the winter, the population would continue to exist at a steady density. On the other hand, if a population is held stable at its normal low density throughout the growing season and then survives winter at the rate of 3.3 percent, the overwintered population at the start of the next year would be abnormally low. However, the winter survival rate of a population at its normal low density may be significantly higher than 3.3 percent.

As suggested earlier in relation to *M. croceipes* releases, two options for keeping greatly reduced overwintered populations below the level of significant damage could be considered. One is to release enough parasites to prevent host populations from increasing. Currently, however, we have no information on the efficiency of parasites when *Heliothis* populations are abnormally low. The second option is to release moths treated for optimum inherited sterility effects. A given ratio of released moths to native moths should have essentially the same influence on reproduction irrespective of the number of native insects in the population or the volume of host-plant material. We will consider in greater detail the potential advantage of integrating the release of parasites and genetically altered moths for regulating *Heliothis* populations in subsequent sections.

General Conclusions

This appraisal indicates that in certain areas Eb may be more useful than *M. croceipes* and *A. marmoratus* for suppressing *Heliothis* populations by the augmentation technique. As compared with these other parasites, Eb parasitizes fewer host larvae but produces a greater number of adult progeny per parasitized host. The greatest advantage of Eb over the other candidates may be its rearing cost. Estimates made by W.C. Nettles (personal communication) indicate that Eb might eventually be mass-produced at a cost comparable to the cost of rearing screwworm flies. The current cost of rearing that pest is less than $1 per 1,000. If overwintered *Heliothis* populations in a typical southern State are as high as 100 million and if a parasite-to-host ratio of 7.5:1 results in adequate suppression, about 750 million Eb would be required. At a cost of $1 per 1,000, the cost for the parasite would be only $750,000. More conservative estimates will be made, however. We will assume a cost of $2 per 1,000 and the need to release parasites for 2 cycles rather than 1. The cost for the parasite would then be about $3 million per State. Therefore, there is reason to believe that Eb offers outstanding potential for suppressing *Heliothis* populations at a small fraction of the cost of damage they cause under current management practices.

The theoretical appraisals of parasites for regulating *Heliothis* populations indicate that at least three species offer promise. Each species may have certain advantages or disadvantages, depending on the ecological area, nature of the host plants, and other factors. Therefore two or more species may have to be released in each area to ensure dependable results. We are dealing with a pest complex that exists on diverse host plants and under widely different ecological conditions. We can be reasonably certain that there will be no simple way of solving this major pest problem.

The release of two or more species not only may ensure more dependable results, because of ecological reasons, but can also result in the enhanced effectiveness of the released parasites. In discussing the release of parasites to suppress the oriental fruit fly (chapter 6), it was shown that the release of *B. longicaudatus* parasites would theoretically increase the rate of parasitsm by the natural population of *B. arisanus*. A model will not be presented, but if both *M. croceipes* and Eb parasites were released during one *Heliothis* generation, the effectiveness of *M. croceipes* would theoretically be greatly enhanced during the second generation if its survival rate were not reduced by superparasitism due to Eb. If enough of each species were released to result in 80 percent parasitism of second- to fourth-instar larvae by *M. croceipes* and 80 percent parasitism of fifth- to sixth-instar larvae by Eb without influencing the survival of the *M. croceipes* parasites, the ratio of adult *M. croceipes* to adult *Heliothis* during generation 2 would theoretically be 20:1. On the other hand, if *M. croceipes* alone were released at a rate that resulted in 80 percent parasitism, the ratio during generation 2 would be only 4:1. The progeny produced by larvae parasitized by Eb would also contribute to suppression during generation 2.

Other opportunities are available for gaining maximum benefits from parasite releases. Insect parasites having a long preoviposition or prelarviposition period are likely to suffer a rather high rate of mortality during that period. Keeping such parasites in confinement until they are ready to attack the hosts might minimize this mortality to the degree that the average host-parasitization capability of the females would be doubled.

Dynamics of *Heliothis* Populations in the United States

Much thought has been given to ways of making meaningful estimates of absolute numbers of *Heliothis* that normally overwinter in the United States and how the populations grow and spread in damaging numbers throughout most of the Nation by season's end. Good estimates are particularly important for *H. zea*. This species is responsible for much of the losses due to *Heliothis*, especially in predominantly corn ecosystems. At present, the only practical means of controlling *H. zea* on field corn is to grow varieties that have considerable resistance to damage the pest causes. Field corn is so vital to the dynamics of *H. zea* populations that this pest would probably be of minor importance on all crops if corn varieties that are near 100 percent effective in inhibiting *H. zea* reproduction were to be developed and grown in all areas.

Since the actual numbers of adult *H. zea* present throughout the Untied States are not known, they must be estimated by indirect methods. The model shown in table 28 shows estimates of the

Table 28
Estimated sizes and trends of *Heliothis zea* populations in southern and northern regions in the United States

Heliothis generation	Adult moth populations (billions)[1]			
	Southern region		Northern region	
	Moths, number	Increase rate	Moths, number	Increase rate
1 (overwintering)	1.5	4	0	0
2	4.5	4	1.5	5
3	13.5	4	12.0	5
4	40.5	overwintering	73.5	overwintering

[1]The population in the northern region is assumed to originate from immigrant moths from the southern, overwintering areas. Twenty-five percent of the moths emerging each generation (except generation 1) are assumed to migrate into the northern region. This would reduce the net increase rate in the South by 25 percent but would increase the number reproducing in the northern region. For example, 1.5 billion of the 6.0 billion moths assumed to emerge in the southern region during generation 2 are expected to emigrate to the northern region. The increase rate is assumed to be 5.0-fold in the northern region. Thus, 1.5 billion would increase to 7.5 billion for generation 3, but this number would be supplemented by 25 percent (4.5 billion) of the 18 billion emerging in the southern region during generation 3. Therefore, the total population would be 12.0 billion in the northern region during generation 3. The number remaining in the southern region would be 13.5 billion. The numbers would increase substantially during generation 4.

yearly numbers and increase rates of *H. zea* populations in the overwintering areas and in the northern areas of spread. A number of variations of the model were developed and analyzed before the one presented was selected. Making such estimates may seem largely academic, especially since there is currently no direct way of determining their accuracy. However, if serious consideration is to be given to suppressing *Heliothis* populations by using density-dependent techniques that involve genetic manipulation, sex pheromones, and parasite releases, it will be important to have some indication of the size of the populations during periods of lowest densities. It will also be important to have some indication of the dynamics of the populations so that the degree of control needed may be estimated.

While there is no direct way of relating the estimated numbers of adult *H. zea* in natural populations to the numbers in the model, there is considerable information on egg and larval densities on wild and cultivated crops. When the total acreages of various crops are related to the fecundity of the adults and to the egg and larval densities that have been recorded, we can use indirect methods to judge whether the estimates are reasonable. The model depicts discrete generations but only for convenience. The generations will overlap in both regions, and the adults move considerably. Therefore, the growth curves would be more gradual than the numbers suggest.

From the standpoint of total population management, the most important estimate is the size of the overwintered population. The ultimate objective would be to prevent the usual growth of populations not only to protect crops in the normal overwintering areas but also to prevent the spread of economically damaging numbers of moths into non-overwintering areas.

In estimating the growth rate of *Heliothis* populations, I have occasionally assumed an average increase of 5.0-fold per generation under conditions favorable for reproduction. However, as discussed earlier, when related to four or more generations per season, this average increase rate is obviously too high, at least for the southern region. At a 5.0-fold increase rate per cycle the seasonal growth rate during 4 cycles would be 625-fold. In such case the egg and larval densities on crops during periods corresponding to generations 3 and 4 would greatly exceed the densities commonly observed. As previously noted, the rate of increase is more likely to average on the order of 2.5-fold per generation. But *H. zea* has characteristics that alter the dynamics of populations during the different generations. When *H. zea* larvae develop in ears of corn, they are well protected from biological agents and have a high probability of survival. For the population in the northern region, I assume a 5-fold rate of increase for generations 3 and 4. Much of the reproduction will be in corn ears, and since the populations are expanding into new areas as the season advances (Snow and Copeland 1971), some of the parasites and other biological agents that are endemic in the overwintering areas probably are not able to keep pace with the *Heliothis* immigrants.

The assumption that 25 percent of the *H. zea* emerging in the southern region emigrate to the northern region each generation except generation 1 is a mere guess. The proportion that emigrates may be higher or lower, but the estimate would seem compatible with the other assumptions. We can make a gross determination as to whether the model seems realistic by relating the estimated adult populations to probable egg and larval densities on the kinds and amounts of host crops in the northern and southern regions.

For example, if 25 billion of the approximately 37 billion females assumed to be present in the northern region during generation 4 each deposit an average of 400 eggs on corn, the total egg population on corn would be 10,000 billion. These are staggering numbers; but assuming that 40 million acres of corn are susceptible to attack at that time, the egg density would average 250,000 per acre, or about 10 eggs per ear of corn, during a period corresponding to 1 generation. The assumed 12 billion remaining females could account for an average of 100,000 eggs per acre on a similar acreage of soybeans. I believe that these estimates are not likely to deviate greatly from actual numbers and that they can account for the heavy losses caused by *H. zea* in the northern region each year.

In a similar manner, we can relate the estimated adult populations in the southern region to egg densities on crops in this region. If 5.0 billion of the approximately 7.0 billion females assumed to be present during generation 3 each deposit an average of 400 eggs on corn in the ear stage during this period, the egg population on corn would total 2,000 billion. Assuming 5 million acres of corn during this period, the egg population would average 400,000 per acre, or an average near 16 eggs per ear of corn. If the approximately 20 billion females assumed to be present during generation 4 each deposit an average of 400 eggs, the total egg population would be about 8,000 billion. This would be enough eggs to result in an average of 160,000 per acre on 50 million acres of cotton, soybean, tobacco, and other crops.

The estimates of egg densities in the South can be roughly compared with egg or larval densities recorded on crops during crop monitoring activities. Puterka et al. (1985) recorded an average of 101,129 *Heliothis* eggs per hectare of corn in the Rolling Plains area in Texas during 1981 and 1982. The estimate of 400,000 eggs per acre, or about 1 million per hectare, during a period corresponding to 1 generation would probably mean an egg population of 100,000 per hectare on any given survey date (based on the assumption that the incubation period is 3 days and that the number of unhatched eggs present on any given survey date is approximately 10 percent of the number present during a 30-day period). The estimate of 400,000 eggs per acre on corn during 1 generation would therefore be in line with the data obtained by Puertka et al. The theoretical estimates of egg densities on other crops during generation 4 do not agree as well with the data recorded by Puertka et al. The estimated average of 160,000 eggs per acre on other crops during 1 generation would mean an average egg density of about 16,000 per acre on a given survey date, or about 40,000 per hectare. This estimate is somewhat higher than the observed egg densities of 12,964 per hectare on cotton and 14,826 per hectare on potatoes as recorded by Puertka et al. These discrepancies suggest, however, that the gross estimates in the model are more likely to be too high than too low.

The estimated numbers of adult *H. zea* in the model can be related to probable actual numbers in other ways. If we assume that an average of 4,000 adult *H. zea* will emerge per acre of cornfields when reproduction is in corn ears (in all probability, a realistic estimate) and that generation 3 will reproduce on 5 million acres of corn in the southern region, the *H. zea* population produced on corn alone would total 20 billion. This number would account for about half the total population estimated for the southern region during generation 4.

It would be desirable to have more precise information on actual numbers of various insect pests that exist in natural populations. Also, refinements in population models are needed. But since such information is so difficult to obtain and yet so vital to an appraisal of the potential of insect-pest-management strategies that are influenced by pest density, appraisal by the indirect procedure described seems to be the only alternative we now have. I have confidence, however, that the indirect estimates made for *H. zea* are not likely to deviate by a wide margin from typical numbers in the two regions.

To further analyze the feasibility of integrating the augmentation and autocidal techniques to suppress *Heliothis* populations, I make the assumption that typically the overwintered populations of *H. zea* and *H. virescens* in the United States are, respectively, on the order of 1.5 billion and 0.5 billion. The *Heliothis* populations during generation 1 in the United States are likely to consist of overwintered moths as well as immigrant moths that developed earlier in Mexico. Hartstack et al. (1982) and Raulston et al. (1982) discussed the evidence of long-distance movement of *Heliothis*. A.N. Sparks (U.S. Department of Agriculture, retired) observed that high populations of *H. zea* develop on corn in northeastern Mexico and that many of the moths probably migrate to the United States (personal communication). Efficient management of *Heliothis* populations may therefore require prior suppression in critical areas in Mexico. The proposed techniques for suppression in the United States should also be applicable in Mexico.

Since the success of *H. zea* management in the United States may depend to a considerable extent on prior management of populations in Mexico, brief comments on the problem in Mexico are offered. We might assume that 500,000 acres of corn grown in northern Mexico could be the source of hundreds of millions of *H. zea* in the United States.

The integrated releases in Mexico of *A. marmoratus* and irradiated moths as proposed for corn-growing areas in the overwintering areas in the United States might well prevent significant movement of fertile *H. zea* into the United States. I propose the release of 1,000 adult *A. marmoratus* per acre during the whorl and tassel stages of corn growth and the subsequent release, per acre, of an average of 500 males treated for inherited sterility effects to further inhibit reproduction of the reduced population. Since the released males would be partially fertile, some reproduction in ears of corn would occur. However, most of the progeny of the treated males mated with normal females would be sterile, as previously noted. If the ratio of released males to native males and females is on the order of 10:1 the ability of the *Heliothis* population to produce fertile progeny would be greatly reduced. About 80 percent of the male and female moths that emigrate to the United States might be sterile. Based on an assumed cost of $10 per 1,000 *A. marmoratus* and a similar cost for the irradiated moths of both sexes, the cost for the insects on 500,000 acres of corn would be $20 per acre, or $10 million. Other costs would also be involved. However, a joint investment of $15–20 million by the United States and Mexico each year might minimize losses to corn growers in Mexico and possibly prevent the movement of hundreds of millions of fertile *H. zea* into the United States early in the season. The immigrant moths from Mexico might alone increase to many billions in the United States before season's end if suppressive measures were not undertaken in the United States.

Scenario of a Program To Regulate Populations of *Heliothis* Species in the United States

A program that I envision might eventually be feasible for maintaining populations of *Heliothis* spp. below damaging levels on all crops in the United States is presented here, and it is based on the estimates and assumptions made previously in this chapter. The program would be comprehensive in scope and demanding in its execution. It would have to be guided by extensive surveys of adult and larval populations and by professionals who know the merits and limitations of the various technologies that might be employed. The cost initially may seem high, but if the program is successful, the continuing annual costs are likely to be a mere fraction of the economic losses that can be expected under present management practices. Since the methods of suppression would be *Heliothis* specific, the benefits from the standpoint of environmental quality may be considered by many to be equally as important as the economic benefits.

Both species of *Heliothis* would be involved, but they will be considered a single entity. Effective management of only one of the species, for example, *H. virescens*, would not, in my opinion, be acceptable to growers. To be completely successful, the program would have to include both species in all of the overwintering areas. There is evidence, as noted, that *H. zea* in particular has the capability of dispersing for hundreds of miles early in the season. Indeed, large numbers of *H. zea* disperse several hundred miles each generation during the growing season. There is no other explanation for the usual spread of *H. zea* throughout the northern region each year. In view of such mobility, it is questionable whether suppression efforts limited to individual States or even to several contiguous States would ensure a high degree of control. As already noted, a highly successful program may also require the cooperation of Mexico and prior suppression in critical crop-growing areas in Mexico before its initiation in the United States.

Eradication of *Heliothis* populations cannot be considered biologically or technically feasible. The alternative would be to maintain populations in the overwintering areas below the level of significant economic damage on all crops. This action should automatically protect the northern, non-overwintering areas.

When viewed from a broad perspective, the envisioned program would involve fulfilling three objectives. The first objective—to be fulfilled during year 1—would be to reduce the overwintered population to a level that could be managed at minimal costs during year 2 and subsequent years by genetic manipulations or possibly by the release of parasites. During year 1, therefore, enough of one or more of the three species *M. croceipes*, *A. marmoratus*, and *E. bryani*, or perhaps other parasite species, would be reared and released to prevent the normal growth of *Heliothis* populations. Genetically altered moths may also have to be reared and then released in some areas during year 1 to supplement the control by the parasites. If the populations during year 1 were merely held stable, natural control factors during the winter, which are probably largely independent of density, should reduce the overwintering populations by 95 percent or more. In such event, as few as 100 million moths out of a normal total of 2 billion overwintering moths of both species would likely survive the winter.

The second objective—to be fulfilled during year 2—would be to maintain suppression of *H. zea* by the release of adults receiving appropriate dosages or irradiation for optimum inherited sterility effects. The suppression of *H. virescens* might be maintained by the release of partially irradiated adults also, although the release of *H. virescens-H. subflexa* hybrid may be preferable. If the populations could be effectively managed during year 2, virtually all economic losses should be prevented during that year in both the normal overwintering region and the northern region. However, a secondary objective of the program during year 2 would be to ensure that the number of moths surviving winter will be smaller in year 3 than it was in year 2. If the estimated population of 100 million adults that survived year 1 were held stable during year 2, natural mortality during the winter might be expected to enable only 10 million or fewer moths to begin year 3 in the United States.

The third objective—to be fulfilled during year 3 and subsequent years—would be to prevent the normal growth of the greatly reduced overwintered population and to also prevent successful reproduction of long-range immigrant moths from sources outside the management areas.

General Discussion

Managing *Heliothis* populations according to the procedures outlined may be viewed by many as beyond practical realization. The program is based on theories, estimates, and assumptions that cannot now be verified. A number of questions for which we do not now have answers can be legitimately raised.

I believe, however, that the basic concept of the program is sound and that realistic parameters have been used to estimate the normal numerical relationships between populations of *M. croceipes*, *A. marmoratus*, *E. bryani*, and their *Heliothis* hosts, and the normal rates of parasitism. The estimated effects of augmented parasite populations are, however, based on extrapolations that extend far beyond any direct supporting evidence. Therefore, the only way to determine whether the projected effects are realistic is to undertake trial programs. But such programs should be on a scale that will virtually eliminate parasite and host movement as a significant variable factor. I would not accept results of trials in nonisolated areas—even on the scale of one State—to be indicative of the influence parasite or moth releases would have on the dynamics of *Heliothis* populations in a closed ecosystem. We are dealing with pests that are known to disperse for hundreds of miles during a single season. The parasites also may move for many miles each year. The parasite augmentation technique as proposed relies not only on the released parasites but also on their progeny produced in subsequent cycles. The assumed ability of the parasites to search for hosts moving from crop to crop and habitat to habitat is considered one of the most important of the behavioral characteristics likely to ensure success of the augmentation technique, provided it is used correctly.

While the release of parasites alone might well reduce *Heliothis* populations below economically damaging levels, the release of genetically altered moths would seem to offer the most practical means for subsequent continuous management. If natural populations are reduced to very low levels by parasites, relatively few moths would be required to achieve very high ratios of released moths to native moths. Excellent detection systems using the sex pheromones would permit the ratios in various parts of the ecosystem to be monitored, and release rates could be adjusted as required. The envisioned program would manage but not eradicate *Heliothis* populations. It would also take full advantage of natural control factors. If very low populations are not permitted to increase in the usual manner, natural mortality each winter would be a major factor in the success of the program.

A very sensitive subject should be addressed, namely, the ecological consequences of rigidly suppressing insect pests. It is a biological axiom that strong and continuous suppression of insect pest populations will adversely affect the biological organisms that are totally or largely dependent on the pests for reproduction and survival. This adverse effect will be the consequence regardless of the way the suppression is achieved. As regards *Heliothis* populations, suppression might be accomplished by an as-yet-to-be-found biological organism that, developing on its own, will maintain the populations below levels of significant damage. Suppression by such an organism would, no doubt, be hailed as a major triumph for the classical biological control approach. Suppression might also be accomplished through the production of crop varieties that are near 100 percent effective in preventing *Heliothis* reproduction. The development of such varieties for all important crops has been a major goal of entomologists and plant breeders for many years. Regardless of the suppression technique employed, if it is target-pest specific, its adverse effects should be minimal compared with techniques that require the repeated use of broad-spectrum insecticides.

Cost Estimates

Estimating the magnitude of the costs that might be involved in a nationwide *Heliothis* management program would, of course, be difficult. However, the estimates of natural population densities and numbers of insects that would have to be released to provide effective suppression have no practical significance unless they can be related to the magnitude of the costs that might be involved in the execution of the program and what the benefits would be. Without doubt, the costs would be high; and administrators, budget officials, and the agricultural interests will no doubt be reluctant to commit the resources required to develop such a program unless the prospects are good that the benefits would substantially exceed the costs. Therefore, some comments seem desirable on the magnitude of the costs and benefits that might be involved.

The estimate of a normal population of 2 billion overwintered *Heliothis* is perhaps the most important single parameter. Table 24 suggests that a 3:1 ratio of adult *M. croceipes* to adult *Heliothis* would result in a high degree of suppression in areas where corn is not a major early-season host plant. Also proposed is the release of *A. marmoratus* in sufficient numbers in corn-growing areas to achieve a 3:1 ratio of adult parasites to adult *H. zea*. For the purpose of this analysis, I assume that 6 billion *M. croceipes* and 3 billion *A. marmoratus* would be required during year 1. While mass-production methods have not been perfected for either parasite species, I would consider it reasonable to assume eventual costs of $20 per 1,000 *M. croceipes* and $10 per 1,000 *A. marmoratus*. In such event, the cost for the parasites during year 1 would be on the order of $150 million. If all other costs during year 1 amounted to $50 million, the total cost during the first year would be on the order of $200 million. *E. bryani* may be the most effective and the most economical parasite to release in some areas.

To some colleagues such a program may seem too costly to even be considered for future development. But we should keep in mind that we are dealing with a pest complex that causes losses on the order of $1 billion per year. Therefore, if *Heliothis* populations were merely stabilized the first year, the benefits may exceed the costs by severalfold. However, the full benefits of the program would not be realized until year 2 and afterward. The release of genetically altered moths might be the most effective and practical procedure during year 2 and subsequent years, but the possibility that continuing releases of appropriate parasite species would be more practical should not be discounted. If overwintered populations were reduced to 100 million by year 2, the release of enough genetically altered moths to achieve a 25:1 ratio of released moths to native moths should provide reasonable assurance that the pest populations would not increase. If genetically altered adult *Heliothis* could eventually be mass-produced at a cost of $10 per

1,000, the release of 2.5 billion such moths during year 2 would cost on the order of $25 million.

If the population during year 2 were merely held stable, virtually all economic losses should be prevented during that year, and natural mortality should reduce the year-3 population to an even lower level. The routine release of as few as 1 billion moths should result in an overall ratio near 100:1. A very high overall ratio, no doubt, would be required to ensure an adequate ratio in critical areas, especially if there is a minimum investment in survey activities. The objective would be to prevent the development of threatening populations from overwintered survivors and to limit reproduction of moths entering the United States from the south. The costs of all aspects of the program during year 3 and afterwards may total less than $25 to $50 million per year. Such a total would represent less than 5 percent of the cost of losses caused by the *Heliothis* spp. under present management practices; moreover, the procedures proposed would be target-pest specific.

The insect-pest-management community has been engaged in intensive research on the *Heliothis* complex for decades and has amassed much information on all aspects of the *Heliothis* problem. It is my conviction that the basic technology necessary to solve the problem satisfactorily is already available. The greatest remaining challenge is to perfect and then apply the technology in a manner that can achieve the objective. The approach described involving total population management would seem worthy of serious consideration. The alternative will be continuing high losses and reactive control measures applied on a farm-to-farm and crop-to-crop basis. The use of fast-acting broad-spectrum insecticides may then be required indefinitely.

10 Suppression of Boll Weevil Populations

Overview

Few insect pests have caused greater losses to agriculture than has the boll weevil, *Anthonomous grandis* Boheman. This obnoxious pest has caused losses totaling many billions of dollars since it entered the United States from Mexico in about 1892 (Coker 1976). Much of the insecticides used for agricultural purposes in the United States can be attributed directly or indirectly to the boll weevil. The intensive use of broad-spectrum insecticides for boll weevil control causes ecological imbalances and intensifies other cotton-insect and crop-pest problems. Thus, in evaluating the importance of the boll weevil under current management practices, consideration must be given to its linkage with other agricultural pests, particularly the *Heliothis* spp.

Cotton growers and many scientists have long hoped that the boll weevil might eventually be eradicated from the United States or, at least, be rigidly managed by ecologically acceptable means so that agricultural losses and environmental harm may be prevented. Consequently, the development of such means has been the focus of much of the research effort during the past 25 years. Scientists have gained a good understanding of the biology, ecology, and dynamics of the boll weevil. Many highly effective insecticides have been developed (Parencia et al. 1983); and efficient ways to use them have also been developed (Brazzel 1959, Brazzel et al. 1961, Lloyd et al. 1966, Rummel 1976). An outstanding achievement was the identification and synthesis of the compounds constituting the boll weevil sex attractant (Tumlinson et al. 1971). This achievement has enabled scientists to develop boll weevil detection methods of unprecedented sensitivity, as well as a means of controlling low level populations (Lloyd et al. 1981, Lloyd et al. 1983). Entomologists and plant breeders have collaborated to develop cotton varieties having fruiting characteristics that minimize boll weevil damage (Namken et al. 1983). Rigid cultural measures provide an effective means of reducing boll weevil population (Summy et al. 1988). Efficient methods of mass-rearing boll weevils at reasonable costs have been devised (Griffin et al. 1983). These methods have made possible the use of the sterility technique for suppressing or eliminating low density populations (Davich 1976, Wright and

Villavaso 1983). The ability to mass-produce the host insect may also be the key to rearing the parasites for use in augmentation programs.

The basic technology for boll weevil population suppression was tested, while still under development, in a boll weevil eradication experiment conducted in Mississippi and Louisiana in 1971–73 and in a trial program conducted in North Carolina during 1978–80. The results of these trials were discussed by various authors contributing to a U.S. Department of Agriculture publication (1976b) and to a publication edited by Ridgway et al. (1983). According to Knipling (1983), the basic technology has undeniably advanced to the point that isolated boll weevil populations can be eradicated or that unisolated populations can be rigidly managed on an areawide basis. Eradication programs currently under way in North and South Carolina and in Arizona have been highly successful (Brazzel et al., manuscript in preparation).

Regardless of the techniques that may be available for eradicating or rigidly managing boll weevil populations, they will have to be used against total populations in a thorough and fully coordinated manner and on a large scale to give the desired results. The boll weevil is a dynamic pest having the potential to increase from very low levels to damaging levels within two or three generational cycles. Some years ago, I calculated that if 10 percent of the cotton acreage is not appropriately treated and the conditions are favorable to boll weevils, enough boll weevils can develop to require that all other growers in the community continue insecticide treatments to prevent excessive damage. Despite the losses they cause and the outstanding advances that have been made to control them, boll weevils have generally been dealt with in an uncoordinated manner for decades.

The negligence of a small proportion of growers, together with the termination of control measures by all growers after the cotton has matured, permits enough boll weevils to develop and overwinter and then cause essentially the same threat to the cotton crop year after year. Many millions of pounds of environmentally hazardous insecticides were used unnecessarily during the past 50 years because of the failure of the pest-management community and the growers to clearly recognize, accept, and advocate the fully coordinated approach to boll weevil population management. In my view, however, the greatest weakness in current boll weevil management technology is the necessity of relying on broad-spectrum insecticides to reduce populations to a level that can be eradicated or continuously managed by ecologically safe means, such as the use of sterile boll weevils or pheromone-baited traps. The use of appropriate parasite species in augmentation programs may offer the possibility to reduce boll weevil populations to low levels without the use of insecticides. For this reason, the boll weevil is included in this appraisal of the potential of the parasite augmentation technique as a means of suppressing insect pest populations. I also wished to show that the boll weevil—one of the major insect pests that are noted for their resistance to control by natural biological agents—will, in theory, be highly susceptible to control by the parasite augmentation technique.

Natural Biological Control Agents

The boll weevil is generally considered invulnerable to effective control by natural biological agents. In this respect, however, it does not differ materially from many other major pest species. Its response to natural control agents is consistent with the hypothesis that regulating forces inherent in nature limit the number of such agents that can develop naturally. The boll weevil, however, is attacked by a number of parasites and other biological agents. Cross and Chestnut (1971) and Cate (1985, and personal communications) listed a wide range of parasites that are endemic in the United States, Mexico, and Central America and that parasitize the boll weevil in natural habitats or in the laboratory. Therefore, several species might be considered for augmentation purposes. *Bracon mellitor* Say is the most prevalent species attacking the boll weevil in the United States. Parasitism data involving this native parasite have been obtained by a number of authors (Pierce 1908, Miller and Crisfield 1930, Adams et al. 1969, Chestnut and Cross 1971, Bottrell 1976, and others). The parasite is widespread and, no doubt, contributes significantly to regulating boll weevil populations. Developing on its own, however, it seems to be a minor factor in reducing damage by the pest. The extensive use of insecticides also limits its role as a natural control agent.

Adams et al. (1969) investigated the biology and behavior of *B. mellitor* and conducted small-scale field-release experiments that led to fairly high rates of parasitism. It can be reared readily in the laboratory. The species attacks a wide range of hosts (Cross and Chestnut 1971), so it is unlikely to preferentially seek out boll weevils, especially when boll weevil densities are low. However, an analysis of *B. mellitor* is presented, and it should serve as a suitable model for the analysis of other species that are similar to *B. mellitor* in behavior but are primary boll weevil parasites.

Considering the recorded rates of parasitism from a seasonal standpoint and the fact that uncontrolled boll weevil populations usually increase manifold from normal-low to normal-high levels during the growing season, there appears to be relatively little positive correlation between boll weevil densities and the rates of parasitism caused by *B. mellitor*. Such a lack of positive correlation is consistent with parasitism data from other parasite-host associations.

Natural Coexistence Pattern of Boll Weevil and *Bracon mellitor* Populations

As developed for other parasite-host complexes, a population model is first developed that is representative of the dynamics of the host population under normal conditions. Estimates are required for three key parameters: (1) the number of adult hosts per unit area during the seasonal generations, (2) the average number of larvae that develop to the stage preferred for parasitization and that are produced per female host each generation, and (3) the rate of survival of the unparasitized hosts to the adult stage each generation. Correct values for these parameters will indicate the rate of growth of the host population during the season.

The model shown in table 29 is proposed as representative of a typical boll weevil population. It will be used to estimate (1) the number of adult *B. mellitor* likely to coexist with the boll weevil population, (2) the average number of host larvae that are parasitized during the lifetime of coexisting female parasites each generation, and (3) the rates of survival of the immature parasites to the adult stage. A basic assumption is that the rates at which the immature parasites and unparasitized host larvae survive to the respective adult stages are similar because they are exposed to many of the same environmental hazards. The comparable survival theory for solitary parasites is the basis for estimating the ratio of adult parasites to adult hosts likely to coexist in natural populations. This ratio can be estimated from the rate of parasitism that is typical for the species under consideration.

While variable, the rates of parasitism recorded for *B. mellitor* indicate that 10 to 15 percent would be reasonably representative of the average rate of parasitism. However, the use of insecticides will reduce the average, as suggested by data obtained by Chestnut and Cross (1971).

Table 29
Population model representing the dynamics of a typical boll weevil population

Parameter[1] (1 acre)	Generation		
	1	2	3
Cotton fruiting forms (squares and bolls)[2]	125,000	250,000	125,000
Female boll weevils	100	600	2,916
Large larvae produced per female (3d–5th instar)	60	54	50
Total large larvae	6,000	32,400	145,800
Squares infested, percent	4.8	13	100
Survival of larvae to adults, percent	20	18	15
Adult boll weevils, next cycle	1,200	5,832	[3]21,870
Increase rate of boll weevil population	6.0	4.9	3.75

[1]Arbitrary reductions in the number of larvae produced per female boll weevil and in the rate of survival are made because of the increasing host density.
[2]Cotton having the fruiting characteristics indicated could be expected to yield about 1 bale per acre in the absence of the boll weevil. A boll weevil population of the nature depicted would have little influence on fruiting during generations 1 and 2 but would destroy virtually all fruiting forms during generation 3. Severe damage to small bolls set during generation 2 would also be expected during generation 3. The yield of untreated cotton would probably be reduced by 1/3 or more, in agreement with data obtained by Parencia and Cowen (1972).
[3]Extensive migration of the boll weevil progeny developing during generation 3 would be expected during generation 4, due to the diminishing number of fruiting forms when the cotton crop matures.

Miller and Crisfield (1930) reported that parasitism due to *B. mellitor* in Georgia averaged 10 percent in June, 14 percent in July, and 18 percent in August. The percentage increased to 35 percent in September. However, late in the season, the amount of fruit is generally limited, and boll weevil movement is extensive. The relationship of parasitism to boll weevil densities would be difficult to estimate under such circumstances. Insecticides were not generally applied when the parasitism data by Miller and Crisfield were obtained. Therefore, I consider their data for June to August to be representative of normal parasitism trends caused by *B. mellitor*. The model shown in table 30 is proposed as representative of the coexistence pattern of a typical boll weevil population and an associated *B. mellitor* population. The most important parameters for the parasites are the estimates of the average host-finding capability of *B. mellitor* females in natural populations, and the proportion of the immature parasites that survive to the adult stage.

The coexistence model is based on the assumption that the ratio of adult parasites to adult boll weevils is 1:9 during generation 1. The

Table 30
Basic model depicting the estimated coexistence of boll weevil and *Bracon mellitor* popultions in a natural environment (see table 29 for boll weevil population model in the absence of the parasite population)

Parameter (1 acre)	Generation		
	1	2	3
Cotton fruiting forms (squares and bolls)	125,000	250,000	125,000
Female boll weevils	100	540	2,316
Large larvae produced per female (3d–5th instar)	60	55	52
Total large larvae	6,000	29,700	120,432
Squares infested, percent	4.8	11.9	96
Female *B. mellitor*	11	61	357
Ratio, adult parasites to adult hosts	1:9	1:8.9	1:6.5
Host larvae parasitized per female	55	65	60
Total host larvae parasitized	605	3,965	21,420
Parasitism, percent	10	13.3	17.8
Host larvae not parasitized	5,395	25,735	99,012
Survival to adults, percent	20	18	16
Adult boll weevils, next generation	1,079	4,632	15,842
Adult *B. mellitor*, next generation	121	714	3,427
Increase rate of host population	5.4	4.3	3.4
Increase rate of parasite population	5.5	5.9	4.8

average number of host larvae parasitized by the female parasite is estimated to be approximately 10 percent lower than the average number of host larvae produced by the female boll weevil. Thus, the rate of parasitism would be 10 percent during generation 1. If the survival rates are equal, as assumed, the ratio of adult parasites to adult boll weevils would be expected to be near 1:9 during generation 2. However, data indicate that the rate of parasitism normally increases to about 14 percent during generation 2 (Miller and Crisfield 1930). To account for this increase, it is assumed that the average number of host larvae produced by the female boll weevil tends to decline slightly as the host density increases and that the average number of host larvae parasitized by the female parasite tends to increase slightly. Possible reasons for the increase in rate of parasitism are (1) the host-finding capability increases because of the increase in host density; (2) parasites searching for hosts on other crops move into the cotton fields to search, supplementing the parasites already present; and (3) the survival rate of the adult boll weevils is lower than the survival rate of the adult parasites, resulting in an increase in the parasite-to-host ratio. If the increase in parasitism is a response to the increase in boll weevil density, the correlation must be very weak. The boll weevil population increases by about 20-fold over three generations, but the rate of parasitism increases from 10 percent to only about 18 percent. It is my conclusion that within the normal boll weevil density range, the average host-finding capability of *B. mellitor* females remains reasonably constant. As discussed in chapter 1, in terms of the average number of adult progeny produced per female parent, the reproductive success of a parasite dependent on a given host for reproduction tends to be comparable to the reproductive success of the host population.

It would largely be fortuitous, if values for the various parameters prove to be highly accurate. But the estimated values could not deviate much from the actual values and still depict the usual parasitism trend indicated by host collecting. Also, the parasite could not survive if it were much less efficient in parasitizing hosts than assumed. Models will not be shown, but if the average number of hosts parasitized per female were only half that estimated and all the other parameter values were held constant, the rates of parasitism would be approximately 5, 3, and 2 percent, respectively, for the three successive generations. On the other hand, if the average efficiency were 50 percent higher than assumed, the rates of parasitism would be about 15, 30, and 70 percent, respectively, during the three generations. In either case, the parasitism trends would not be consistent with available data. The true efficiency must fall between these estimates. A clue to the host-finding efficiency of parasites that I find useful is the fecundity of the species. According to Adams et al. (1969), the *B. mellitor* female produces an average of about 164 eggs under laboratory conditions. I generally assume that due to natural mortality during the adult stage, the actual number of eggs deposited is near 33.3 percent of the fecundity. On this basis, the average would be approximately 55 and would be in line with the estimates in the model. I have confidence that the actual host-finding capability of *B. mellitor* under conditions favorable for reproduction deviates by less than 25 percent from the efficiency assumed. I also have confidence that the relative values assigned to the other parameters are approximately correct. If they are, the coexistence model should be suitable as a basis for calculating, by extrapolation, the most likely effects of parasite releases on the rates of parasitism and on the dynamics of boll weevil populations when the releases are made throughout boll weevil ecosystems during the first reproducing generation.

Table 30 suggests that self-perpetuating populations of *B. mellitor* would have some influence on the dynamics of coexisting boll weevil populations but would not prevent excessive damage. Virtually all of the fruiting forms would be completely destroyed by late in the season. As indicated in table 29 (footnote), the damage to squares and bolls would likely reduce cotton yields by about one-third. I believe that the basic models presented (tables 29 and 30) realistically represent the dynamics of typical, uncontrolled boll weevil populations and of typical, naturally coexisting populations of boll weevils and *B. mellitor.* The key parameter values presented are considered sufficiently accurate to make a good estimate of the influence that releases of *B. mellitor* or similar parasites would have in closed boll weevil ecosystems.

Influence of *Catolaccus grandi* Releases

The estimate of the host-parasitization capability of *Bracon mellitor*, a prevalent parasite that is native to the United States, can be used to calculate the influence of other similar parasites on the dynamics of boll weevil populations when released into boll weevil habitats. Perhaps more effective than *B. mellitor* for augmentation purposes is *Catolaccus* (= *Heterolaccus*) *grandi* Burks, a primary boll weevil parasite that is similar to *B. mellitor* in behavior. *C. grandi* is not known to exist in the United States, but it is reported to be widespread in Mexico and Central America. Limited parasitism data reported by Cross and Mitchell (1969) indicate that the natural parasitism rate of *C. grandi* is rather low, on the order of 10 percent in some locations. But, as discussed elsewhere in this report, a parasite's natural rate of parasitism is not necessarily indicative of its inherent efficiency in parasitizing its host or of its potential effectiveness in augmentation programs. The rate merely reflects the relative numbers of the adult parasites and adult hosts that can normally coexist as a result of natural balancing mechanisms.

Johnson et al. (1973) investigated the biology of *C. grandi.* It parasitizes the larger boll weevil larvae (3d instars) in squares and bolls, and was reared successfully in the laboratory on infested squares. Its life cycle seems to be well synchronized with that of the boll weevil. Host insects other than the boll weevil are not known to exist.

Johnson et al. released the parasite in 1/10-acre plots of cotton in Mississippi. It reproduced, and in one series of releases, the estimated number of adult progeny produced by season's end exceeded the number released by 5-fold. There is no way of knowing, however, how many of the released parasites or their progeny may have left the small test areas. Therefore, the number of progeny may have far exceeded the observed 5-fold growth rate. In my opinion, the results obtained from the releases in the small plots are very promising. There is no way of determining the growth rate of a released population or the rate of parasitism caused by released parasites and their progeny unless the releases are made in large areas or against isolated boll weevil populations. As a number of hypothetical models indicate, the progeny produced from cycle to cycle can be expected to contribute more to control than the parasites released. In my view, *C. grandi* would be an excellent species to consider for suppressing boll weevil populations by the augmentation procedure. Apparently, it is not able to survive the winters. This, however, would not be an important limiting factor in augmentation programs if the parasite reproduces normally during the growing season, as indicated by the release experiments referred to.

For this theoretical appraisal, it is assumed that *C. grandi* has a host-finding capability similar to the estimated capability of *B. mellitor* and that the boll weevil is the only available host. The boll weevil population model depicted in table 29 will be used to estimate the influence of *C. grandi* releases. The suppression model is shown in table 31. Enough *C. grandi* adults are released during generation 1 to result in a 2.5:1 ratio of adult parasites to adult hosts. Assuming a population of 100 female boll weevils per acre, the release of 250 female *C. grandi* per acre would be required. Higher or lower natural boll weevil populations would require proportionally more or fewer parasites. The female *C. grandi* is estimated to find an average of 55 host larvae during its lifetime under normal conditions. However, in view of the abnormally high population created by the releases and the expected higher-than-normal predation upon the parasite, its average efficiency is estimated to be reduced by 20 percent. Based on these estimates, the adult female would parasitize an average of 44 boll weevil larvae during the first boll weevil generation. The parasites are assumed to search independently.

The various parameter values and the method used to calculate the results are indicated in table 31. The rate of parasitism resulting from the released parasites would not be high enough during generation 1 to appreciably reduce the boll weevil population. However a parasitization rate of 84 percent during generation 1 would theoretically result in a 5.2:1 ratio of adult parasites to adult hosts during generation 2, or about two times the initial ratio. The higher parasite population would likely sustain a higher rate of predation; hence, the estimated average host-finding capability during generation 2 is assumed to be reduced by an additional 20 percent. Despite this adjustment, the ratio of parasite-host encounters to hosts present would increase substantially. The 3:1

Table 31
Suppression model depicting the estimated influence of *Catolaccus grandi* releases on the rate of larval parasitism and on the dynamics of a boll weevil population (see table 29 for the basic boll weevil population model)

Parameter[1] (1 acre)	Generation			
	1	2	3	[2]4
Cotton fruiting forms (squares and bolls)	125,000	250,000	125,000	—
Female boll weevils	100	96	29	2.5
Large larvae produced per female	60	60	70	75
Total large larvae	6,000	5,760	2,030	188
Percent squares infested	4.8	2.3	1.6	—
Female *C. grandi* (released during generation 1)	250	504	547	251
Ratio, adult parasites to adult hosts	2.5:1	5.2:1	19:1	99:1
Host larvae parasitized per female parasite[3]	44	35	30	—
Total parasite-host encounters	11,000	17,640	16,410	—
Ratio, parasite-host encounters to hosts present[4]	1.8:1	3:1	8:1	—
Parasitism, percent	84	95	99	—
Host larvae parasitized, number	5,040	5,472	2,010	—
Host larvae not parasitized	960	288	20	—
Survival rates to adults, percent	20	20	25	—
Adult boll weevils, next generation	192	58	5	—
Adult *C. grandis*, next generation	1,008	1,094	502	—

[1]When the host density decreases significantly, the average number of host larvae produced and the rate of survival are assumed to increase.
[2]Because of the abnormally low host density, no effort is made to estimate the values for many of the parameters in generation 4, including the rate of parasitism. However, the ratio of adult parasites to adult hosts would theoretically be 99:1. If the host population numbers 188 larvae and each female parasite parasitized 4 larvae, the ratio of parasite-host encounters to hosts present would exceed 5:1, and parasitism would again reach 99 percent.
[3]In view of the abnormally high parasite population, the average host parasitization efficiency of the females is assumed to decrease by 20 percent during generation 1. A further reduction of 20 percent is assumed during generation 2 because of the increase in the parasite population. A further reduction in the average host larvae parasitized is assumed during generation 3 because of the lower host density.
[4]See table 1 to calculate the rates of parasitism and escape from parasitism by using the ratio of parasite-host encounters to hosts present.

ratio would theoretically result in 95 percent parasitism during generation 2. If the survival rates of the immature parasites and unparasitized host larvae are equal, the parasite-to-host ratio should be 19:1 during generation 3. In that generation, the larval host population would theoretically be only about 67 percent lower than that in generation 1, because of the high intrinsic rate of increase of the boll weevil population. Whether or not a 67-percent reduction in host larvae would cause a reduction in the average host-finding capability is questionable, but I assume a reduction to 30 host larvae per female. Even so, the ratio of parasites to boll weevils would be so high that the ratio of parasite-host encounters to hosts present would theoretically be 8:1, and parasitism would theoretically exceed 99 percent. The boll weevil population would be reduced to five per acre during generation 4, and the ratio of *C. grandi* to adult boll weevils would be 99:1. Even if the average number of larvae produced by the female boll weevil were to increase to 75 and the *C. grandi* female were to parasitize an average of only 4 larvae because of the low boll weevil density, the theoretical rate of parasitism would still be about 99 percent.

Discussion

A logical question would be, Are the high rates of parasitism depicted in the model realistic regardless of the number of *C. grandi* or any other species that might be present in a boll weevil ecoystem? I have some reservations that a released parasite population would in fact multiply to the extent that it would virtually eliminate the host population. Yet, I believe that realistic and, perhaps, even conservative allowances were made for the biological phenomena that might be expected to become involved under the abnormal conditions created by artificial means. The model is therefore presented for consideration and analysis by authorities on the biological control of the boll weevil.

Comments on some of the factors that are likely to become involved would seem desirable. Perhaps we know at least as much about absolute numbers of boll weevils in natural populations as about any other insect pest. Therefore, I feel that the basic model reasonably represents, in numerical terms, a typical boll weevil population and how it will grow under average conditions. The addition of 250 *C. grandi* per acre during the first seasonal cycle would result in a very high ratio of parasites to hosts. It must be kept in mind that the model applies to closed ecosystems in which all the acreages of cotton would have the assumed 2.5:1 ratio of parasites to boll weevils. This ratio is approximately 22 times the estimated natural ratio, which is assumed to result in 10 percent parasitism. Therefore, the 84-percent rate of parasitism estimated for generation 1 would seem reasonable. Any difference between the survival rates of the immature parasites and unparasitized host larvae to the respective adult stages is not likely to be large enough to significantly alter the estimated ratio of *C. grandi* to boll weevils during generation 2. The ratio would be near 5:1, or about two times the ratio during generation 1. The boll weevil population would not change significantly in size because of its high intrinsic rate of increase. Therefore, in generation 2, the density of the boll weevil should have little or no effect on the efficiency of the parasites to find hosts. The *C. grandi* population, however, would change considerably, namely, by a factor of 2. The assumption that the higher population will result in 20 percent higher predation and, hence, a 20-percent further reduction in the average host-parasitization capability of the female would seem reasonable. Despite such a reduction, a higher proportion of boll weevil larvae would be parasitized during generation 2 than during generation 1. Therefore, the ratio of *C. grandi* to boll weevils would increase further during generation 3.

Whether the estimated 67-percent decline in the larval population from generation 1 to generation 3 would significantly reduce the host-finding capability of the female parasites is doubtful, but even if this capability were only half the assumed capability, parasitism would still approach 99 percent, and the ratio of *C. grandi* to boll weevils would theoretically increase to about 99:1 during generation 4.

When the various assumptions on which the model is based are viewed from the standpoint of relative numbers, it would be difficult to envision how the boll weevil population could escape near elimination. By generation 3, the ratio of *C. grandi* to boll weevils would theoretically exceed a normal ratio by more than 100-fold. But the boll weevil population would still be fairly high, so the rate of parasitism might well be near 99 percent. We will never know, however, what impact such a high ratio will have on a boll weevil population until the augmentation technique is tried in an appropriate manner. Even if the maximum rate of parasitism were on the order of 95 percent during generations 2 to 4 and only 10 percent of the overwintering boll weevils were to survive the winter, the population during year 2 should be reduced to less than one boll weevil per acre. Several ecologically acceptable options may then be available to further suppress or continuously manage the population. These options might involve the release of sterile boll weevils, the use of the boll weevil sex pheromone in traps or as an attracticide; or, the continued releases of a suitable parasite in relatively low numbers. Regardless of the technique that might be employed, however, it would have to be directed against the total population or applied in an area larger than the flight range of the adult boll weevils and parasites.

Cost Estimates

As previously stated, the theoretical results of insect population suppression techniques are of little practical significance unless they can be related to costs likely to be involved in the practical application of the techniques. Boll weevils can be reared in unlimited numbers at reasonable cost. Mass-rearing methods have not been developed for boll weevil parasites. Cate (1987), however, has made considerable progress in rearing *C. grandi* and other boll weevil parasites. There is reason, therefore, to believe that entomologists and engineers could develop methods of mass-producing *C. grandi* (or some other suitable species) at reasonable costs. In my opinion, *C. grandi* could eventually be mass-produced at a cost of $20 per 1,000. The suppression model (table 31) indicates that the release of an average of 500 male and female parasites per acre during generation 1 would adequately control a typical boll weevil population. However, in actual practice, we might assume that this number be released for two successive generations. Therefore, the cost per acre for the parasites would be about $20. Releases should be made reasonably proportional to the number of boll weevils present in the various fields of cotton included in the augmentation program. Therefore, thorough surveys would be a vital aspect of such a program. The total cost, however, should be substantially less than the cost of reducing populations to low levels by the use of insecticides.

These rough estimates suggest that there would be justification for fully investigating promising candidate parasite species for use in augmentation programs to greatly suppress boll weevil populations.

Release of Parasites To Suppress Boll Weevil Populations in Marginal Survival Areas

There may be several million acres of cotton grown in areas where boll weevils normally exist at low levels because of severe winter weather and/or limited favorable hibernating sites. Such areas may be categorized as marginal survival areas. Economic losses caused by the boll weevil in such areas are generally sporadic, and the benefits do not usually justify the cost of control. However, if programs are undertaken to eliminate boll weevil populations from certain areas, it would also be necessary to eliminate populations in adjacent marginal survival areas. Doubtless, these latter populations persist largely because of localized favorable habitats and/or because of immigration from adjacent high-density areas. In theory, if immigration did not occur and if merely enough suppressive pressure were applied in the marginal survival areas to prevent the normal seasonal growth of populations, natural control factors in the winter would, within 2 or 3 years, ensure the virtual disappearance of the boll weevil from those areas.

From the standpoint of individual growers, the cost of eradicating boll weevils from marginal survival areas may not be justified, especially if the eradication measures involved the use of insecticides that might intensify other cotton-insect problems, such as those due to *Heliothis* spp. Therefore, low-cost control procedures that do not lead to ecological imbalances are needed. While the sterility and sex attractant techniques could contribute to this goal, localized populations may often be too high for their use. The suppression model shown in table 31 suggests that the release of a minimum number of parasites might prove to be the most effective strategy to employ in controlling boll weevil populations in marginal survival areas. This possibility will be considered from a theoretical viewpoint.

As previously noted, information is lacking on the influence of abnormally low host densities on the searching behavior and host-finding capability of parasites that are dependent on a given host for survival. However, I have advanced arguments from time to time in this report to support the hypothesis that due to their highly developed host-habitat detection mechanism, host-specific parasites have the inherent capability of finding and parasitizing enough hosts to maintain viable populations even when the host densities are abnormally low. This hypothesis would seem to be particularly applicable to the boll weevil. Female boll weevils generally deposit their eggs in limited areas. The "clumping" of boll weevil infestations is well recognized by boll weevil biologists. The larval progeny of a single female may be concentrated in no more than 1,000 square feet of cotton plantings. Thus, if populations average five females per acre, for example, clumps or patches of host larvae are likely to be concentrated in a total of 5,000 square feet of cotton plants and about 40,000 square feet would be host free. The distribution pattern of the host larvae in the clumps is not likely to differ materially from the distribution pattern on all of the cotton if 44 females per acre were reproducing. Therefore, if the parasites can readily locate the scattered host clumps, the probability of a female parasite finding a given number of hosts during its lifetime may be about the same whether the adult female boll weevil population averages 5 or 44 per acre. Theoretically, the maximum difference in the average host-finding capability at these two host densities would depend on the relative amounts of host-searching time that the parasites spend in the nonhost habitats versus the host habitats. As discussed elsewhere in this report, investigators have observed that certain parasites devote much less time in nonhost habitats than in habitats harboring hosts. This behavior is, no doubt, a characteristic of all parasites. If the ratio of time spent in host habitats versus the time spent in nonhost habitats is 10:1 but the ratio of the area harboring hosts to the area that is host free is approximately 1:8, as it would be if five females existed per acre, the female parasites should find about 56 percent as many hosts at the low density as at the higher density during a given host-searching period. Even this comparison may not indicate the relative average numbers of hosts that would be found during the lifetime of the parasites at the two densities. When host densities are very low, the parasites may merely have to search for hosts more intensively or for longer periods each day to maximize the number of hosts parasitized. But granting that the host-finding ability at a host density of 5 females per acre will be 56 percent of that at a host density of 44 females per acre, only one-fifth as many parasites would be required for the desired suppressive effect at the lower host density.

While this analysis of host-seeking capability at abnormally low host densities is largely theoretical, it is consistent with observations that for many parasites no clear positive correlation has been established between rate of parasitism and host density, even though host densities may vary by up to 100-fold.

Suppression Model

The *C. grandi* female is estimated to be capable of finding an average of 55 host larvae during its lifetime when the boll weevil population averages 100 females per acre. For this appraisal of the feasibility of releasing parasites to control the boll weevil in marginal survival areas, we will assume that the adult boll weevil population averages 5 females per acre and that each female parasite will find an estimated average of 20 host larvae. This latter estimate may be very conservative. Based on models not shown, I calculated that *C. grandi* populations would be incapable of surviving for long if the average host-finding capability were lower than 20 hosts per female during its lifetime. If the average should be as low as 10 hosts per female at this low boll weevil density, it would be necessary to release two times as many parasites to achieve the rate of parasitism estimated for the first generation.

The model shown in table 32 indicates the rates of parasitism and the influence on the dynamics of the boll weevil population that may be expected if 100 *C. grandi* of both sexes are released per acre when the boll weevil population averages 10 of both sexes per acre. The sexes are assumed to be equal in numbers. If the

Table 32
Suppression model depicting the estimated influence of releasing enough *Catolaccus grandi* to exceed an abnormally low boll weevil population by a ratio of 10:1

Parameter (1 acre)	Generation	
	[1]1	[1]2
Female boll weevils	5	1.5
Host larvae per female	60	65
Total host larvae	300	98
Female *C. grandi*, added generation 1	50	36
Ratio, *C. grandi*, to boll weevil	10:1	24:1
Host larvae parasitized per female	20	10
Total parasite-host encounters	1,000	360
Ratio parasite-host encounters to host larvae present[2]	3.3:1	3.7:1
Parasitism, percent	96	97
Host larvae parasitized	288	95
Host larvae not parasitized	12	3
Survival rates to adults, percent	25	30
Boll weevils, next generation	3	1
C. grandi, next generation	72	28

[1]Based on the estimated number and the survival rate of larvae produced by the female boll weevil, the potential increase rate of the population would be 7.5-fold during generation 1 and about 10-fold during generation 2.
[2]To calculate rates of parasitism and escape from parasitism by using the ratio of parasite-host encounters to hosts present, see data in table 1.

estimates are realistic, the boll weevil population would be virtually eliminated during the second generation.

The cost of such a method of suppressing low boll weevil populations would be minimal if the rearing costs for the parasite were on the order of $20 per 1,000. In such event, the cost of 100 *C. grandi* of both sexes would be $2 per acre. Since actual results are not likely to be equal to the theoretical, releases for two generations might be assumed. Thus, the cost for the parasites would be $4 per acre. If costs for releases, surveys, and other activities amounted to $6 per acre, the total cost would be $10 per acre. The model suggests that the population would be virtually eliminated during the first year. However, if releases merely prevented the normal growth in the boll weevil population during year 1, they would still contribute greatly to control, since the survival rate during the winter in marginal survival areas would likely be no higher than 5 percent. In such case, the average density would be less than one boll weevil per acre at the start of year 2. The program during year 2 may largely involve detection and elimination of threatening infestations. It is my opinion that during the many years that boll weevils have been subjected to intensive control, we have not taken full advantage of the forces of nature in devising strategies for regulating or eradicating their populations. In many areas, populations are not permitted to increase significantly during most of the cotton-growing period. But then control measures are terminated, allowing one or two generations to develop normally before hibernation time. If populations were not permitted to grow significantly throughout the season, natural control during the winter would likely result in a further reduction of 90 to 95 percent. The appraisal described may be a realistic estimate of the requirements for eliminating low-level boll weevil populations by the parasite augmentation technique.

Discussion

A critical analysis of the various estimates and assumptions forming the basis of the boll weevil suppression model in table 32 indicates that most of the estimates and assumptions cannot be verified for lack or insufficiency of data. Therefore, the natural reaction will be to question the validity of the model. Since it will be difficult and costly to determine whether the theoretical results are valid, administrators, growers, and scientists themselves may be reluctant to commit resources to make that determination. We might therefore profit by considering the proposed approach from a broad perspective to decide by simple rationalization whether the prospects seem sufficiently favorable to justify committing resources for the determination.

Achieving the parasite-to-host ratios depicted in table 32 should be well within the range of technical feasibility. Boll weevil densities on the order of 10 per acre often occur naturally or can be readily reduced to such a level. We know, within reasonable margins of error, how many parasites would have to be released to achieve a given parasite-to-host ratio. The relative numbers of *C. grandi* and boll weevils in the hypothetical model can, therefore, be considered realistic projections. The basic assumption of equal survival rates for the immature parasites and unparasitized host larvae to the respective adult stages, as previously noted, is not likely to be in error by a wide margin. Therefore, these parameter values can also be regarded as realistic. If parasitism due to *C. grandi* averages 10 percent in its native environment, and the boll weevil population averages five males and five females per acre, the average density of the female *C. grandi* population should be no more than one female per 2 acres of cotton. The suppression model is based on the assumption that enough *C. grandi* will be released to raise the density to 50 females per acre. Thus, we would be increasing the expected normal ratio by about 100-fold. It must be envisioned that this distorted average numerical relationship would exist in every field of cotton in an area larger than the dispersal range of the parasite and the boll weevil.

Host-specific parasites are highly specialized organisms that have demonstrated, through many millennia, their capability of finding enough hosts to survive under a wide range of circumstances. It would not be unreasonable to assume that one female *C. grandi* is more capable of detecting the whereabouts of a reproducing female boll weevil in a 1-acre or a 10-acre cottonfield than any human using the best detection methods we have devised. If the assumption is correct, then it may not be unreasonable to also assume that 50 female parasites on 1 acre are more capable of destroying the boll weevils more effectively and more efficiently than we can with the best methods of control we now have.

Disregarding the many questions that might be raised as to the accuracy of some of the assumptions on which the hypothetical model is based, if we consider the inherent behavior of the parasites and the magnitude of change in numerical relationship between the parasite and boll weevil populations that could be achieved by artificial means, the possibility would seem excellent that the model presented depicts the general magnitude of the results that could be achieved. Considering the need for low-cost and ecologically acceptable approaches to boll weevil suppression, there would seem to be ample justification for research scientists to thoroughly investigate the concept of boll weevil population suppression in the manner proposed. I predict high risks of failure or disappointing results, however, unless the parasites are released in adequate numbers and in areas large enough to virtually eliminate the effects of significant movement of the released parasites and/or their progeny from the release area and of boll weevils into the release area.

Prospects

In the future, the cotton industry may have few if any highly effective insecticides to rely on for controlling the boll weevil in severely affected areas. The insecticide resistance problem has not yet been solved. The public is increasingly apprehensive over environmental hazards created by non-pest-specific insecticides. Thus, there will be greater restrictions on the use of insecticides. While there may be questions as to the validity of some of the estimates and assumptions on which this appraisal of the boll weevil is based, I have confidence that they are biologically tenable. Under present circumstances, I would urge that trial programs be undertaken using the theoretical models as guides.

11 Summary, Conclusions and Practical Implications

A theoretical appraisal was made of the role certain parasite species play as natural control agents for some of the more important insect pests. A primary objective was to estimate, in numerical terms, the average host-parasitization capability of the female parasites in natural populations during their lifetime when conditions for reproduction are normal. Based on the estimates, calculations were made of the number and proportion of a host population that will be parasitized when the parasite population is increased by artificial means throughout the pest-host ecosystem.

Information is not available on actual or relative numbers of parasites and hosts that normally coexist in natural populations, because obtaining this information has been one of the most neglected aspects of research on insect population ecology. These numbers therefore had to be estimated, largely by theoretical deductive procedures, for the various parasite-host associations evaluated. The ratio of naturally coexisting adult parasites to adult hosts largely determines the proportion of the host population that will be parasitized.

The general procedure to appraise the potential of a parasite as a natural control agent was to develop a model that is considered to reliably represent the numbers and growth trends of the host and parasite populations during their annual or generational cycles. The first step was to estimate three key parameters for the host population: (1) the number of adults that exist in a specified area each cycle, (2) the average number of parasite-vulnerable immature stages (eggs or larvae) produced by the female host during its lifetime, and (3) the rate of survival of the immature unparasitized hosts to the adult stage. Appropriate values for the parameters will indicate the rate of growth of the host population during the various cycles.

The next step was to estimate the coexistence pattern of the associated parasite species. Information on the biology and behavior and the rates of parasitism caused by the species were the only useful clues in assigning values for the following parameters: (1) the number of adult parasites coexisting with the host population, (2) the average number of hosts that will be parasitized during the life of the female parasite, and (3) the rate at which the immature parasites in the parasitized hosts survive to the adult stage.

While very simple and based on only a few parameters, the models used greatly facilitated the theoretical investigation. It would be difficult to recognize and fully appreciate the influence of some of the factors involved in parasitism without the use of such models. A number of models based on different assumed values for the various parameters were usually developed before selection was made of the models that are regarded as the most representative of the normal coexistence patterns of the different parasite-host associations. The factors that determine and influence the coexistence pattern of host and associated parasite populations are discussed in considerable detail in chapters 2 and 3. The host and parasite species evaluated are listed in chapter 1.

The parasite species evaluated differ in biology and parasitizing behavior, but they all complete their development in the larval or early pupal stage of the host. Two species (*Chelonus insularius* and *Biosteres arisanus*) parasitize the host eggs and complete development in the host larvae. *Microcentrus grandii* is a polyembryonic species, and *Eucelatoria bryani* has multiparasitism behavior. Several progeny of these last two species generally mature in a single host larva. The other species are solitary larval (or larval-pupal) parasites; and generally, only one parasite completes development in one parasitized host.

A number of theories are proposed to explain how the insect parasitization process works in natural environments. Based on the various theories, estimates were made of the influence of augmented parasite populations on the rates of parasitism and on the dynamics of the host population following the release of the parasites for one or more host cycles. The results obtained are discussed in chapters 4–10. Each parasite-host association has evolved certain unique characteristics that govern the coexistence pattern of the two organisms and the role the parasite species can play as a natural control agent. These are discussed in the various chapters dealing with the different hosts and the parasites evaluated. Parasite-host relationships are governed by the actions and interaction of a number of complex biological phenomena that must be critically analyzed and also evaluated for their significance.

In the sections that follow, brief summaries are presented of the conclusions reached on the parasitization processes in natural parasite-host populations, the conclusions reached on the effects of artificially augmenting parasite populations, and the implications of the theoretical results obtained in this investigation.

Factors Governing Natural Parasitism

1. While many factors are involved in various degrees, the most important single factor that determines the rate of parasitism in natural populations is the relative numbers (ratio) of adult parasites and adult hosts that normally coexist in an environment. For the parasite-host associations evaluated, the host density has little or no influence on the relative numbers of parasites and hosts that coexist in natural populations. Rather, environmental factors influencing the two organisms govern the relative numbers of parasites and hosts that coexist.

2. During the period that the host and parasite populations increase from their normal low to their normal peak densities, the reproductive successes of the two organisms tend to be equal in terms of the average number of adult progeny produced by the female parents. This tendency also holds for some species during periods of declining populations. Thus, for many parasite-host complexes, the ratio of parasites to hosts and the rates of parasitism tend to remain reasonably constant, even though the host populations may fluctuate by manifold. However, for any host whose population usually collapses after reaching a peak, the collapse is accompanied by a slower decline in parasite population. Thus, the ratio of adult parasites to adult hosts and the rate of parasitism will temporarily increase. Such an increase in rate of parasitism cannot be attributed to a prior positive response to the host density. Indeed, there is generally a reduction in the average number of hosts parasitized per female parasite during the period the host population is growing.

3. The population of a host and that of its associated dependent parasite cannot remain in balance long if the females of the two populations differ significantly in reproductive success, as expressed in terms of the average number of adult progeny produced per female during its lifetime. If more than one parasite species depends on a given host for survival, the reproductive success of each of the species cannot be significantly different from that of the host. This equality in reproductive success is true regardless of whether the females of the competing species differ in parasitization behavior or average

rate of parasitization. The relative rates of parasitism caused by the species are not a reflection of the species' inherent host-searching or host-finding capabilities. They are merely a reflection of the relative numbers of parasites and hosts that are present in the environment. Parasite species are often described as "good" or "poor" host searchers and host finders, depending on the relative rates of parasitism they cause. Host-specific parasites are highly specialized organisms. We can assume that they are all excellent host searchers and host finders at all host densities. Otherwise, they could not have evolved and survived. This assumption has practical implications in the selection of a species for augmentation purposes if the costs of mass-producing several appropriate candidate species differ markedly.

4. As regards rate of survival to adulthood, the rate for the host stage or stages preferred for parasitization by solitary larval parasites included in this study and the rate for the immature parasites developing in the hosts are very closely linked. In several respects, the host and host-dependent parasite live as a single entity. Any hazard, either biotic or abiotic, that destroys the host before the developing parasite matures will automatically destroy the immature parasite. The mortality of the pupal stage of the host and that of the pupal stage of its associated parasite are also likely to be closely linked, even though the two pupal stages may live apart. While some mortalities may occur independent of host or parasite mortality, their total is considered minor compared with the total mortality the two organisms share. Therefore, as a working hypothesis, the survival rates are assumed to be equal for most of the species complexes evaluated; and the hypothesis applies at all host densities. Thus, for solitary larval parasites coexisting with their hosts, the ratio of adult parasites to adult hosts is approximately equal to the ratio of parasitized hosts to unparasitized hosts during the previous generation. If parasitism averages 10 percent, the ratio will be near 1:9; if parasitism averages 25 percent, the ratio will be near 1:3, and so forth. As applied to solitary parasites, this hypothesis is considered one of the most important foundations of this study. It makes possible the development of models that reflect reasonably accurate estimates of the normal numerical relationship between the members of various parasite-host associations, as judged on the basis of available data on the normal rates of parasitism. This hypothesis is also important for estimating the rate of parasitism resulting from an increase

or decrease in the ratio of adult parasites to adult hosts in coexisting populations.

If there is a substantial and consistent difference (for example, 25 percent) in the rates of survival of the immature parasites and unparasitized hosts to the respective adult stages, the female parasites must compensate by parasitizing more or fewer hosts relative to the number of hosts produced by the coexisting host females. Otherwise, the two populations would become seriously imbalanced within a few cycles and in time could not coexist. Parasite species that are subjected to high rates of hyperparasitism or that predispose the parasitized hosts to premature death by predation must compensate by parasitizing more hosts. Realistic models that reflect normal population and parasitism trends cannot be developed unless these theories are accepted as valid in principle.

5. The relative average numbers of hosts produced per female host and of hosts parasitized per female parasite tend to be similar; but usually, the average number produced per female host exceeds the average number parasitized by an amount that would ensure the continued survival of the host species. The amount can be estimated, as discussed in chapter 2. Thus, for example, if parasitism by a certain species averages 20 percent, the parasite-to-host ratio would be about 1:4; and the host female during its lifetime will, on average, have to produce 25 percent more host larvae than are parasitized during the lifetime of the coexisting female parasite.

6. The discovery by parasite behaviorists that most, if not all, parasite species possess highly developed mechanisms for detecting host habitats and locating hosts is without question one of the most significant developments in the field of biological control. The ability of parasites to find their hosts by utilizing chemical cues called kairomones means that they have the ability to find near their normal quota of hosts even when the hosts are scarce and widely distributed. I estimate that when the host density is well below the normal low, the average host-finding ability of the females of a host-dependent parasite species is in the order of 10 times higher than the average would be if host searching were completely random. In effect, so long as the host densities are within or near the characteristic density range for the species, the highly developed host-finding capability of parasites virtually eliminates host density as a significant factor influencing the average host-finding

capability of the parasites during their lifetime and the rate of parasitism that the species will achieve. This highly developed capability would explain why no evidence has been found for a positive correlation between host density and rate of parasitism among many parasite-host associations. An equally important possible explanation is that the survival rates of the immature parasites and unparasitized hosts are approximately equal, as noted in item 4.

7. While the highly developed host-finding ability of parasites may explain why host density within the normal range is not a significant factor influencing the normal rate of parasitism, we do not yet know whether this ability is highly enough developed to allow parasites to find hosts effectively at abnormally low host densities. Such densities must doubtless arise from time to time under natural conditions. It would therefore seem logical to assume that the population of any given host species may become so low and so widely dispersed that not enough hosts would be found to maintain a viable parasite population. The size of the area occupied by the two organisms would then become a major factor influencing the host-finding ability of the parasites. Since many parasite species have survived for many thousands of years and, no doubt, have often survived periods of abnormally low host densities, the likelihood is that host populations must decline to very low levels before inability to find enough hosts becomes a critical survival factor. Chemical cues, such as kairomones, that guide parasites to the host habitats and hosts are, no doubt, as vital for parasite survival as certain other insect-produced chemical cues are for insect survival. Sex pheromones produced by insects, for example, enable the species to readily find mates for successful reproduction. Indirect evidence indicates that insect populations, in general, must decline to very low levels before their survival is in jeopardy because of inability to find enough mates. The continued existence of many host-specific parasite species for eons is likewise indirect evidence that abnormally low host densities from time to time will not seriously jeopardize survival of parasite populations; however, no data are available to quantitatively define a "critically low host density."

8. The theory that the average host-finding ability of parasites is not seriously influenced by normal variations in host densities or that the average host-finding capability of the parasites does not decline proportionally to normal declines in host densities has great practical significance. It means that the most strate-

gic time to release parasites in augmentation programs is when the host densities are low. High parasite-to-host ratios can be achieved with a minimal number of parasites under such conditions. The theory also means that a given number of parasites maintained in a host ecosystem by artificial means will become increasingly effective as the host population declines, because the ratio of parasites to hosts will increase. Thus, like the control techniques involving the use of genetically altered adult pests or the sex pheromones of pest insects, the parasite release technique also increases in effectiveness with decreasing pest density.

9. While natural balancing forces and behavioral characteristics tend to maintain rather consistent numerical relationships between host populations and their associated parasite populations, temporary imbalances of these relationships inevitably arise for various reasons. Imbalances are especially likely to occur in agricultural areas because of changes in cultural practices and the use of pesticides. However, when imbalances occur, the same natural regulating forces that tend to maintain a reasonable balance between all competing organisms will come into play. These forces will act with greater or lesser intensity, as required, on the host and parasite populations to eventually restore the normal balance. If the normal rate of parasitism sustained by certain host species averages 25 percent, a normal parasite-to-host ratio of about 1:3 would be likely. If the ratio is reduced for any reason to 1:9, which would result in about 10 percent parasitism, but the reproductive success of the parasite population is increased by 25 percent over the normal because of the decrease in parasite density, the normal parasite-to-host ratio and the normal rate of parasitism would be restored in about five generations.

10. We know that a below-normal parasite-to-host ratio favors the reproductive success of the parasite. When a new parasite species is introduced to control an insect pest in accordance with classical biological control methods, the host population has usually already reached a high level and may have become reasonably well stabilized with the environment. The overall ratio of parasites to hosts when a parasite species is first introduced and released in inoculative numbers may be no higher than 1:10,000, and only about 0.01 percent of the total host population would be parasitized. However, if the average rate of increase of the parasite population each generation averages 50 percent higher than that of the host population

until the parasite population reaches its steady density, a 1:3 ratio of parasites to hosts would be reached by about the 20th generation and would result in about 25 percent parasitism. By then, the total parasite population would have grown by more than 3,000-fold, but the host population would have remained stable or declined considerably because of the new parasite. Since the mortality of the immature parasite stage is so closely linked to the mortality of the host stage parasitized, the advantage of a low parasite density is largely for the adult stage of the parasite.

If the natural balancing forces did not favor the reproductive success of parasites whenever their populations existed below the characteristic densities, the introduction of a new species for biological control would not result in a period of increasing rate of parasitism, and the theory of the eventual establishment of steady-density parasite and host populations would not be valid.

11. Information on the fecundity, biology, and parasitism behavior of various parasite species provides excellent clues for estimating the average host-finding ability of the females during their lifetime, and for estimating the survival rates of the immature parasites to the adult stage. Virtually no definite information now exists on these two key parameters; yet that information is highly important for a good understanding of the natural numerical relationships between the members of parasite-host associations and for estimating the influence of augmented parasite populations on rates of parasitism and on the dynamics of the host populations. Deductive procedures were therefore used to estimate realistic values for the two key parameters.

    The numbers of eggs or larvae produced by the females of the different parasite species evaluated have no relationship to the relative efficiencies of the parasites as natural control agents. They merely reflect the hazards the progeny normally face before they reach the adult stage. A species capable of depositing many hundreds or several thousands of eggs or larvae per adult female and another species capable of depositing far fewer eggs or larvae per female will be equally successful in producing adult progeny if both species depend on the same host for survival and reproduction. We may assume that the fecundities of the two species are 800 and 100 eggs per female and that the females of both species attack their hosts directly.

Even without substantiating data, we can say that the probable error would be small if the assumption were made that each female of the high-fecundity species will parasitize about eight times the number of hosts parasitized by each female of the low-fecundity species. But a further assumption is that the immature parasites of the latter species will be about eight times more likely to survive to the adult stage as those of the high-fecundity species. Thus, both species will be equally successful in terms of the number of adult progeny produced per female parent, as discussed in item 3.

The species having a fecundity of 800 eggs will, in all probability, parasitize one of the early host stages, such as the eggs; and the immature parasites will complete development when the host larvae reach near maturity. The species having a fecundity of 100 eggs will probably parasitize near-mature host larvae. The parasitization and survival patterns for the two species would then suggest that the normal mortality of the host from the egg stage to the large-larva stage is likely to be on the order of 87.5 percent.

The modeling procedure was used to estimate the normal host-finding efficiencies of the females of a number of parasite species. When the estimates made by this procedure were related to information on the fecundity of the various species that attack their hosts directly, I reached the conclusion that a reasonably good efficiency estimate could also be made by dividing the fecundity value by 3. In effect, this estimate would imply that about 67 percent of the mortality of the potential progeny will occur during the adult parasite stage. No doubt, some parasite species will be more or less likely to sustain this mortality rate as a result of environmental hazards during the adult stage, but I believe that this simple rule-of-thumb estimate is not likely to deviate from the true efficiency by more than 25 percent. If so, such rough estimates would probably be more accurate than estimates made by the best techniques we now have for measuring the efficiency of parasites in natural populations.

Not all parasites attack their host directly. Some species, such as *Archytas marmoratus*, deposit their larvae on the host plants, and the parasite larvae must find the larval hosts. Another species, *Blepharipa pratensis*, deposits its eggs on foliage in gypsy moth larval habitats. The large host larvae must ingest the parasite eggs to become parasitized. Such parasitizing behavior introduces a major hazard not faced by species that attack their hosts directly. *A. marmoratus* and *B. pratensis* females, respectively, are each capable of depositing about 2,000 larvae and 4,000 eggs. Judging from the fecundity of parasites that parasitize similar stages of the host larvae directly, the chances are probably no greater than 5 to 10 percent that the immature *A. marmoratus* and *B. pratensis* stages will be successful in parasitizing their hosts. While these immature stages sustain a high total mortality, the adult females compensate by producing a large number of eggs or larvae.

Of the parasites evaluated in this study, some attack the host directly and others deposit their eggs or larvae in the host environment. Some parasitize the host eggs, and the immature parasites complete their development when the hosts reach the large-larva stage. Others parasitize small, medium-size, or large host larvae. Most of the parasites evaluated are solitary species, but one is polyembryonic. The fecundity of the different species ranges from less than 100 immature stages per female to more than 4,000. While the parasites may parasitize their hosts in different ways, for stabilized populations, they will all be equally successful in terms of the average number of adult progeny produced per female parent. A wide diversity in parasitizing behavior also suggests that during millions of years of coexistence with their hosts, parasites have developed modes of living that will ensure successful reproduction with minimal stress on the welfare of the hosts on which they depend for their existence. The two organisms are engaged in a game of numbers, namely, actual and relative numbers of parasites and hosts that can survive to perpetuate their kind from year to year. The normal numerical relationships that have evolved remain within a very narrow range. The practical implication of these numerical relationships is that we can eventually develop the technology to alter them by manifold. The challenge we face is to determine what influence such alterations will have on the rates at which host populations are parasitized and on how these rates influence the dynamics of the host populations.

Influence of Parasite Releases

1. The parasites are assumed to be released at a strategic time throughout a pest-host ecosystem. The most strategic time is when conditions for reproduction are favorable but the host

population is at or below its normal low level. The objective of the releases is to increase the rate of parasitism enough to maintain total pest-host populations below economically damaging levels throughout the ecosystem. Releasing the parasites in small, unisolated areas would be inconsistent with this objective because the parasites and their progeny would be able to move out of the area and/or pest hosts from neighboring areas would be able to move into the release area. This investigation deals entirely with the concept of total host population management by the augmentation technique. All of the conclusions drawn from this study are based on the premise that the parasites and hosts coexist in closed ecosystems.

2. Although high mobility of parasites and hosts can greatly negate the effectiveness and practicality of the parasite augmentation technique when it is applied to only a small part of a pest-host ecosystem, such mobility is considered highly beneficial when the technique is applied to the entire pest ecosystem. The ability of parasites to readily move from host habitat to host habitat and disperse long distances, if necessary, together with their highly developed host-detection capability, permits the parasites to find their necessary quota of hosts at variable host densities. Parasite mobility is considered particularly advantageous when the pest-host density is low and the area of infestation large. No other method of insect pest control now employed has the characteristic of actively "seeking" and destroying the pest wherever it exists within the area of application. The pest host actually serves as a powerful attractant for the parasite. Whether the host-finding capability of parasites remains undiminished when pest-host populations fall to abnormally low levels remains to be determined. Without question, however, insect parasites, with their high mobility and high capability to find hosts, have the potential of contributing greatly to the effective management and, possibly, eradication of pest populations.

3. The estimated effects of parasite releases are based on the premise that individual female parasites search for hosts independently. If 1 female under certain conditions has the ability to find and parasitize x number of hosts during its lifetime, 10 will parasitize $10x$, 1,000 will parasitize $1{,}000x$, and so forth. However, the average number of hosts that will be successfully parasitized per female parasite will decrease as the parasite population grows and as the rate of parasitism increases.

The higher the parasite population in relation to the host population, the greater will be the rate at which the parasites suffer predation; and the average number of hosts parasitized per female parasite will decrease because the average lifespan of the female parasites will decrease. Arbitrary allowances are made for this factor.

Also, a greatly increasing parasite density will result not only in progressive decreases in the average number of hosts parasitized per female parasite but also, and more importantly, in the parasites encountering an increasing number of previously parasitized hosts. When such a host is encountered, the female may or may not reparasitize it, depending on the discriminatory behavior of the parasite species. For solitary larval parasites, one parasitized host is assumed capable of producing only one parasite progeny. Therefore, all previously parasitized hosts that are found would represent reduced efficiency. As noted in table 1, when parasitism reaches a level of about 86 percent, the laws of chance dictate that the hosts will have been encountered an average of about two times. This frequency would represent a 50-percent loss in parasitization efficiency. When parasitism reaches 95 percent, the hosts will have been encountered an average of about three times. This frequently would represent about 67 percent loss in efficiency. Encountering previously parasitized hosts is not a factor of major importance for natural populations that normally parasitize only 10 to 20 percent of the hosts, because the probability of such encounters will be low; but it is of major importance for augmented populations because the objective is to achieve very high rates of parasitism. For species that do not discriminate, excessive reparasitization may reduce the rate of survival of the immature parasites. This effect on parasite survival is another factor that must be considered in estimating the influence of parasite releases for the various parasite-host complexes, as discussed in chapters 4–10.

4. Since the major objective of this theoretical investigation was to estimate the potential of the augmentation technique in suppressing and maintaining total pest-host populations below levels of economic damage, it is assumed that enough parasites will be released to achieve the suppressive effects desired. The intrinsic rate of increase of many major insect pests is likely to be on the order of 5-fold per generation when conditions for reproduction are favorable. Usually, therefore, the goal will be to release enough parasites to raise the rate of parasitism to

near 80 percent during the host generation that the releases are made.

5. It is estimated that for solitary parasite species, the ratio of adult parasites to adult hosts must be on the order of 2.5:1 to result in 80 percent parasitism. This ratio is assumed to apply to all solitary parasite species, regardless of their normal rates of parasitism in natural environments or the host stage or stages they parasitize. As noted, if natural parasitism averages 10 percent, the normal ratio is likely to be near 1:9; and if natural parasitism averages 25 percent, the normal ratio will be about 1:3. Thus, a ratio of 2.5:1, achieved by releasing the parasites, would be about 22.5 times the ratio that causes 10 percent parasitism and about 7.5 times the ratio that causes 25 percent parasitism. It would seem reasonable to assume that such magnitudes of increase in parasite-to-host ratios will result in the estimated high rates of parasitism shown in the various models during the host generation that the releases are made.

6. The rate of parasitism achieved by the adult progeny produced by the released parasites will largely determine the benefits to be derived from the parasite augmentation technique. Among certain parasite-host associations, the higher the rate of parasitism achieved during the first host cycle, the higher will be the rate of parasitism during the second cycle because of the progeny produced. This continuing increase in rate of parasitism is a characteristic that makes the parasite augmentation technique so unique and potentially effective for managing pest populations. The pest-host's resources are used to maintain or even enhance suppressive action for several successive host cycles without the need to release additional parasites. This characteristic of using the pest-host's resources is also a characteristic of natural, or self-perpetuating, parasite populations. But for many species, the natural rates of parasitism are not high enough to cause satisfactory control, because their parasitizing behavior and natural environmental constraints limit the relative numbers of parasites and hosts that coexist in natural populations.

7. The self-perpetuating populations of biological agents are fundamental to natural biological control. They are equally fundamental to successful biological control by the parasite augmentation technique. However, determining the number of progeny produced by augmented populations has received little

investigation under field conditions. Therefore, virtually no data exist that would allow quantitative estimates to be made of the influence of parasite progeny on host populations. For this reason, no other phase of this investigation received more consideration than the theoretical basis for estimating the number of parasite progeny produced per generation, and the rates of parasitism each generation will achieve.

As previously discussed, the relative survival rates of immature parasites and unparasitized host larvae to the respective adult stages are assumed to be similar for most of the species included in this study. Therefore, as an example, if the rate of parasitism is 80 percent when the ratio of adult parasites to adult hosts is 2.5:1, the ratio of adult parasites to adult hosts in the next cycle should be about 80:20, or 4:1. This ratio is considerably higher than that estimated to cause 80 percent parasitism. A parasitization rate of 80 percent will not cause a reduction in a host population whose intrinsic rate of increase is 5-fold or more. For such a population, therefore, the average host-finding capability of the female parasites should not decline during generation 2, because the host density has not changed. In any event, it is assumed that the average number of hosts parasitized per female parasite does not decline proportionally to the decline in a host population. Such a lack of correlation is to be expected because of the role that kairomones or other host-guidance mechanisms play in host finding.

If a parasite-to-host ratio of 2.5:1 during generation 1 causes 80 percent parasitism, a ratio of 4:1 should cause on the order of 90 percent parasitism during generation 2 after allowances are made for intraspecific competition and changes in host and parasite densities. Then, the ratio of adult parasites to adult hosts would be about 9:1 during generation 3 if the survival rates continue to remain essentially equal. Therefore, an even higher rate of parasitism can be anticipated during generation 3 if, because of the role of kairomones, the average host-finding capability of the females does not decrease proportionally to the decline in the host population.

These estimates are the basis for the conclusion that for certain parasites, an initial, sufficient augmentation of their numbers will lead to a progressive increase in rate of parasitism with each succeeding generation of the parasites. In theory, this progression can continue until the host population is virtually eliminated. This biological action would be comparable to the long-sought meiotic drive mechanism that could lead to virtual extinction of an organism.

8. For the kinds of parasites evaluated, a progressive increase in rate of parasitism as described in item 7 cannot be expected to occur unless the rate of parasitism achieved during generation 1 exceeds 50 percent. If the augmentative release of parasites causes less than 50 percent parasitism, the rate of parasitism during the next few generations is unlikely to increase. Instead, the population density of the parasites will decrease gradually to its steady density. Enough control can be expected, however, to prevent the normal increase of the host population. Any significant increases in rates of parasitism achieved in augmentation programs will have an adverse effect on the dynamics of the pest-host population, and the benefits may exceed the costs. For a high degree of control, however, the general guideline might be to release enough parasites initially to achieve more than 50 percent parasitism.

9. Certain parasite-host complexes have evolved a natural regulating mechanism that prevents a progressive increase in parasite-to-host ratio even when the release of parasites initially causes the rate of parasitism to substantially exceed 50 percent. The mechanism involves an increase in the mortality rate of the superparasitized hosts when the rate of parasitism increases. Two complexes evaluated use this mechanism, and their parasitc members are both egg-larval species. The species are *Biosteres arisanus*, which parasitizes eggs of tropical fruit flies, and *Chelonus insularius*, which parasitizes eggs of noctuids. These parasites complete development in the host larvae. How this mechanism works and its significance are discussed in some detail in chapters 6 and 7. In brief, however, the higher the rate of parasitism of the host eggs, the higher the mortality of the eggs that are parasitized and superparasitized. The result is that even though the rate of parasitism could theoretically approach 100 percent, the rate of parasitism of the host larvae will probably never increase beyond about 75 percent. The ratio of adult parasites to adult hosts will therefore be restricted, because the survival rates of the parasite larvae and unparasitized host larvae to the respective adult stages tend to be equal.

   This mortality factor is considered a natural control mechanism that certain parasite-host associations have evolved to prevent complete domination of host populations by highly efficient parasite species. Theoretically, however, this natural regulating mechanism can be readily nullified by the release each generation of supplemenatal adults to compensate for the mortality caused by superparasitism. Whether a pest-host population can overcome and survive such treatment is questionable, as discussed by use of models in chapters 6 and 7.

10. One of the most important conclusions reached in this investigation is that in nature, conditions can never occur that will lead to the sustained coexistence of abnormally high populations of a host-specific parasite with abnormally low populations of the host. Such a sustained imbalance in parasite and host populations would be totally inconsistent with nature's balancing mechanisms. But another important conclusion reached in this investigation is that it should eventually be technically and economically feasible to artificially create and maintain such an imbalance between the populations of many key parasite species and those of their associated hosts. When the populations of a host insect are low whether because of natural events or the application of suppressive measures, the host-dependent parasite can be released in large enough numbers to ensure for an indefinite number of cycles the establishment of parasite-to-host ratios that will exceed the normal ratios by as much as 10- to 100-fold or more, as required.

    Since such distorted ratios cannot occur naturally and have never been created by artificial means throughout any pest-host ecosystem, their impact on rates of parasitism and on the dynamics of pest populations can only be speculated. But we know that we would be releasing a highly specialized and efficient organism that inherently possesses highly developed host-detection capabilities even when the host populations are low. A logical question can be raised, Are there any biological mechanisms that can come into play with sufficient intensity to prevent a high degree of host suppression or, perhaps, complete elimination if such abnormal ratios are maintained throughout a pest-host ecosystem for a number of successive cyles? The suppression models presented, which are based largely on the factors that govern parasitism processes, indicate that the existence of such mechanisms is unlikely and that virtual elimination of pest-host populations would be inevitable. However, the full impact of such artificially created changes in the numerical relationship between parasite and host populations cannot be determined until appropriate field trials are conducted.

11. The parasite augmentation and autocidal techniques have some similar and some dissimilar suppressive characteristics. The efficiency of both techniques is pest-density dependent. It is generally accepted that the ratio of sterile insects to native insects is directly and linearly related to suppression efficiency. There is strong theoretical evidence that within the range of normal host densities, the ratio of parasites to hosts is also directly and linearly related to suppression efficiency. But whether this relationship becomes invalid after a host population reaches some critical low level is not known. If parasites and fully competitive sterile insects were to be employed against separate populations of the same pest species existing at some normal level, I estimate that a 1:1 ratio of adult parasites to adult hosts and a 1:1 ratio of sterile insects to fertile insects will result in about 50 percent control of both populations. However, I estimate also that as the ratios increase, the parasite augmentation technique will become more effective than the autocidal technique. For example, if a sterile-to-fertile ratio of 1:1 causes 50 percent sterile matings, a ratio of 2:1 should cause 67 percent sterile matings, and a ratio of 9:1 should cause 90 percent sterile matings; however, if a parasite-to-host ratio of 1:1 results in 50 percent parasitism, a ratio of 2:1 should cause 75 percent parasitism, and a ratio as high as 9:1 would theoretically result in more than 99 percent parasitism. Some adjustments may be necessary to allow for increasing parasite density. Still, I estimate that to match the effectiveness of a parasite-to-host ratio as high as 9:1, a sterile-to-fertile ratio of at least 99:1 would be required.

    When genetically altered insects are released, they compete against their fertile counterparts that constitute the native population. Such competition does not exist when parasites are released. In fact, if any native parasites of the species released are present, they will add to the degree of protection. This absence of competition is one of the several important differences in the principles of suppression that favor the parasite release technique.

12. The comparisons made in item 11 apply only for the generation present when the insects are released. Sterile insects released during one pest cycle will have little or no effect during the next cycle. For continued control, additional sterile insects must be

released each cycle, although a constant release rate becomes increasingly effective if the pest population is declining. On the other hand, when parasites are released, not only do they achieve control during the generation that the releases are made, but the parasite progeny can be expected to cause a comparable or even higher degree of control in subsequent generations. When all factors are considered, I estimate that at high ratios, 1 parasite will be as effective as about 25 sterile insects, except possibly when the pest population exists near extinction. We know that a low density does not lower the efficiency of the sterility technique. The sterile insect and parasite augmentation techniques are highly complementary in suppressive action and are also, in theory, synergistic. Therefore, appropriate integration of the two techniques may prove to be highly advantageous, resulting in a very high degree of suppression or elimination of pest populations.

13. To suppress the population of a pest insect, the concurrent release of two parasite species may be more ecologically advantageous than the release of only one species. Also, concurrent releases can result in synergistic parasitism if the species differ in parasitism behavior. In the generations following the releases, the total suppressive action resulting from the concurrent release of two species will theoretically be greater than that resulting from the release of an equal number of parasites of only one of the species. It is for this reason that the rate of parasitism by a natural parasite population can be enhanced by the release of a species that parasitizes a more advanced host stage without significantly reducing the survival rate of the hosts already parasitized. Such enhancement can be expected if a high degree of parasitism by the released species reduces the number of adults in the host population without reducing the number of adults that can develop in the natural parasite population. The result would be an increase in the ratio of adult parasites to adult hosts in the natural population during generation 2, and an increase in the normal rate of parasitism by the natural population. Theoretical models indicate a substantial advantage in releasing a given number of parasites of a species that does not adversely influence the reproduction and survival of a natural population that causes a high degree of parasitism in nature.

14. The rate of parasitism by a natural, host-specific population of parasites can also be enhanced if any pest-specific method of control is applied against the adults of the pest population.

(Examples are autocidal techniques, use of baits containing insect sex attractants, and use of pest-specific biological or chemical insecticides.) The rate of parasitism will be enhanced during the next pest generation because the ratio of adult parasites to adult hosts will be above normal. Target-pest-specific adult-control measures permit the normal survival of not only the parasite population but also the total complex of natural biological agents. Therefore, their use makes theoretically possible a significant bonus effect—the enhanced biological control by all of the natural agents. This effect should be further inducement for the development and use of target-pest-specific methods of insect control.

15. The conclusions reached in this investigation apply to the kinds of parasites and hosts analyzed. All of the parasite-host associations evaluated have evolved somewhat different coexistance patterns. For more gregarious hosts such as aphids, scale insects, and certain coleoptera, a theoretical analysis along lines undertaken in this investigation might indicate parasite-to-host ratios and parasitism trends considerably different from those presented in this report. Also, the relative numbers and parasitism trends are likely to differ considerably from those estimated if the parasites have more or fewer cycles than their hosts.

Practical Implications

The suppression models developed in this investigation indicate that the release of appropriate numbers of parasites in a closed pest-host ecosystem will greatly suppress the pest population—sometimes to near extinction levels—before natural balancing forces can restore the normal parasite-to-host ratio and the normal rates of parasitism. For some species, supplemental parasite releases will be necessary for one or two host cycles to compensate for high mortalities among the superparasitized hosts.

As a general guideline, enough parasites should be released during the first host cycle to cause on the order of 80 percent parasitism. As noted for solitary parasites, this rate of parasitism is estimated to require a parasite-to-host ratio of about 2.5:1. The degree of control during the first season, together with normal winter mortality (or a comparable hostile period for tropical species), is conservatively estimated to reduce the host population to such a low level that during year 2 only about one-tenth the number of parasites released in year 1 would be required to keep the host population low. If the estimate is accurate, the host population during year 3 should decline to an even lower level, and fewer parasites would be required. If the pest population is totally isolated, complete elimination will theoretically be the final result. In areas subject to reinvasion by immigrating hosts, the routine releases of relatively few parasites should prevent the reestablishment of damaging populations. In a general way, this describes the strategy proposed for making optimum use of the parasite augmentation technique.

The release of parasites may also be one of the most practical means of reducing normal, low-density populations of some pests to much lower levels. In many cases it should be practical to achieve parasite-to-host ratios ranging from 10:1 to 100:1. I question whether any pest population could continue to cause significant economic damage if it is subjected to parasite releases that maintain the parasite-to-host ratios as high as 10:1 for a sustained period. I question also whether any host population could even survive for more than a few cycles if the parasite-to-host ratio were maintained as high as 100:1. Parasite releases might also play a prominent role in eliminating newly discovered pest introduction. In such circumstances, it would likely be practical to release parasites in numbers that would exceed the new pest populations by many thousandfold. The possibility of employing such an ecologically acceptable means of eliminating certain insect pests should be given due consideration.

The ultimate cost of parasite releases for suppressing and managing insect populations will depend largely on the progress that entomologists and engineers make in the mass-production of various parasite species. Thus, the development of efficient parasite-rearing technology will be one of the most important aspects of pest population management by the parasite augmentation technique.

While there is reason to have confidence that the various suppression models are based on valid theories and sound biological principles, the models will assuredly have to be validated by well-planned and well-executed release experiments before the results they depict can be accepted as representative of the results to be expected in practical programs. Unfortunately, validation of such models will be a very difficult and costly undertaking for virtually every parasite-host association. Much greater research resources will be required than have ever been made available for investigating the augmentation technique. Well-adapted parasite species will have to be released at the appropriate time, in adequate numbers, and throughout the target pest ecosystem. Thorough surveys will have to be conducted to guide the operational programs. The experimental areas will have to be isolated or large enough to compensate for the mobilities of the insect pests and their associated parasites. An unisolated area encompassing as many as 1,000 square miles may still be too small for some parasite-host complexes. If possible, parasite releases should be made against host populations that are well isolated, such as those on islands or in ecoystems having physical or ecological barriers to the excessive spread of the released parasites and their progeny.

While the costs for appropriate pilot tests will be high, the stakes also are very high, and the rewards are potentially very great. If well-conducted trial programs produce results that merely approach the theoretical results, there would seem to be almost unlimited potential for using the parasite augmentation technique to manage populations of many of the world's major insect pests. Nature has already provided an abundant resource that can be drawn upon for the selection of appropriate parasite species for use in parasite release programs.

Several new mechansisms of pest population suppression were revealed by the theoretical investigation. Of special interest is the synergistic suppressive action that will result when the parasite augmentation and the autocidal techniques are aplied concurrently. In theory, the potentiating actions are of such magnitude that the integration of the two techniques could provide a method of pest population management or eradication of extraordinary efficiency. If enough of a host-specific parasite is released to achieve a parasite-to-host ratio of 2.5:1, and enough sterile insects are also released at the appropriate time to achieve an effective sterile-to-fertile ratio of 4:1, the combined suppression in numerical terms during the first host cycle would represent about a 5-fold enhancement of the suppression that would be caused by the release of only the parasites or only the sterile insects. If the same number of sterile insects is then released during the second cycle, the suppression in numerical terms would be enhanced by about 100-fold. While largely academic, the suppression during cycle 3 would be enhanced by more than 1,000-fold. This powerful suppression mechanism could add a new dimension to pest management strategies for the future.

When the results of the theoretical investigation presented in this report are viewed from a broad perspective, there is reason to conclude that the parasite augmentation technique employed along lines proposed offers the potential for rigidly regulating populations of many major insect pests at costs that might be less than one-tenth the cost of losses the pests cause under current management practices. The potential benefits to U.S. agriculture alone could amount to billions of dollars each year. Additionally, the use of target-pest-specific methods would make a major contribution to improving environmental quality.

Literature Cited

Adams, C.H., W.H. Cross, and H.C. Mitchell. 1969. Biology of *Bracon mellitor*, a parasite of the boll weevil. J. Econ. Entomol. 62:889–895.

Amoako-Atta, B., R.E. Dennell, and R.B. Mills. 1978. Radiation induced sterility in *Ephestia cautella* (Walk) (Lepidoptera: Pyralidae): Recovery of fertility during five generations after gamma irradiation. J. Stored Prod. Res. 14:181.

Arbuthnot, K.D. 1955. European corn borer parasite complex near East Hartford, Connecticut. J. Econ. Entomol. 48:91–3

Arthur, Alfred P. 1981. Host acceptance by parasitoids. *In* D.A. Nordlund, R.L. Jones, and W.J. Lewis (eds.), Semiochemicals: Their Role in Pest Control. pp. 97–120. Wiley, New York, Chichester, Brisbane, Toronto.

Ashley, T.R., C.S. Barfield, V.H. Waddell, and E.R. Mitchell. 1983. Parasitization of fall armyworm larvae on volunteer corn, Bermuda grass, and paragrass. Fla. Entomol. 66:267–271.

Bedford, E.C.G. 1956. The automatic collection of mass reared parasites in two consignment boxes, using two light sources. Entomol. Soc. S. Afr. J. 19:342–353.

Beglyarov, G.A., and A.I. Smetnik. 1977. Seasonal colonization of entomophages in the USSR. *In* R.L. Ridgway and S.B. Vinson (eds.), Biological Control by Augmentation of Natural Enemies. pp. 283–328. Plenum Press, New York and London, 480 pp.

Bell, R.A., C.D. Owens, M. Shapiro, and J.R. Tarditt. 1981. Mass rearing and virus production. *In* The Gypsy Moth: Research Toward Integrated Pest Management. U.S. Dep. Agric. Tech. Bull. 1584.

Bennett, F.D. 1969. Tachinid flies as biological control agents for sugarcane moth borers. *In* J.R. Williams, J.R. Metcalf, R.W. Montgomery, and R. Mathis (eds.), Pests of Sugarcane. pp. 117–148. Elsevier, New York, Amsterdam, London.

Bess, H.A., F.H. Haramoto, and D. Hinckley. 1963. Population studies of the oriental fruit fly, *Dacus dorsalis* Hendel (Diptera: Tephritidae). Ecology 44:197–201.

Bess, H.A., R. vandenBosch, and F.H. Haramoto. 1961. Fruit fly parasites and their activities in Hawaii. Proc. Hawaii. Entomol. Soc. 17:367–378.

Biliotti, E. 1977. Augmentation of natural enemies in western Europe. *In* R.L. Ridgway and S.B. Vinson (eds.), Biological Control by Augmentation of Natural Enemies. pp. 341–377. Plenum Press, New York, London, 480 pp.

Blumenthal, E.M., R.A. Fusco, and R.C. Reardon. 1979. Augmentation release of two established parasite species to suppress populations of the gypsy moth. J. Econ. Entomol. 72:281–288.

Botelho, P.S.M., N. Macedo, and A.C. Mendes. 1980. Aspects of the population dynamics of *Apanteles flavipes* (Cameron) and support capacity of its host, *Diatraea saccharalis* (Fabr.). XVII ISSCT Congress, Manila, The Philippines, pp. 1736–1744.

Bottrell, D.G. 1976. Biological control agents of the boll weevil. *In* Boll Weevil Suppression, Management, and Elimination Technology. U.S. Dep. Agric., Agric. Res. Serv., ARS S–71, pp. 22–25.

Bradley, W.G. 1952. The European corn borer. U.S. Dep. Agric. Yearb. 1952, pp. 614–621.

Braham, R.R., and J.A. Witter. 1978. Consumption of foilage of juvenile and mature red oak trees by late instar gypsy moth larvae. J. Econ. Entomol. 71:425–426.

Brazzel, J.R. 1959. The effect of late season applications of insecticides on diapausing boll weevils. J. Econ. Entomol. 52:1042–1045.

Brazzel, J.R., T.B. Davich, and L.D. Harris. 1961. A new approach to boll weevil control. J. Econ. Entomol. 54:723–730.

Browning, H.W., and E.R. Oatman. 1984. Intra- and interspecific relationships among some parasites of *Trichoplusia ni.* Environ. Entomol. 13:551–556.

Bryan, D.E., C.G. Jackson, and R. Patana. 1970. Biological comparison of the two species of *Eucelatoria* parasites in *Heliothis* spp. J. Econ. Entomol. 63:1469–72.

Bryan, D.E., C.G. Jackson, and R. Patana. 1972. Production of progeny and longevity of *Eucelatoria* sp. parasite in *Heliothis* spp. Environ. Entomol. 1:23–26.

Burleigh, J.G., and J.H. Farmer. 1978. Dynamics of *Heliothis* spp. larval parasitism in Southeast Arkansas. Environ. Entomol. 7:692–694.

Campbell, Robert W. 1981. Population dynamics. *In* The Gypsy Moth: Research Toward Integrated Pest Management. U.S. Dep. Agric., Tech. Bull. 1584, 757 pp.

Carpenter, J.E., J.R. Young, E.F. Knipling, and A.N. Sparks. 1983. Fall armyworm (Lepidoptera: Noctuidae): Inheritance of gamma-induced deleterious effects and potential for pest control. J. Econ. Entomol. 76:378–382.

Cate, J.R. 1985. Cotton: Status and current limitations to biological control in Texas and Arkansas. *In* Marjorie A. Hoy and Donald C. Herzog (eds.), Biological Control in Agricultural IPM Systems. pp. 537–556. Academic Press, New York. 389 pp.

Cate, J.R. 1987. A method of rearing parasitoids of boll weevil without the host plant. Southwest. Entomol. 12:211–215.

Charpentier, L.J., W.J. McCormick, and R. Mathis. 1960. Biological control of the sugarcane borer in Louisiana. Proc. Int. Soc. Sugar Cane Technol. 10:865–9.

Chestnut, T.L., and W.H. Cross. 1971. Arthropod parasites of the boll weevil, *Anthonomus grandis*: 2. Comparison of their importance in the United States over a period of 38 years. Ann. Entomol. Soc. Am. 64:549–557.

Clausen, C.P. 1956. Biological control of insect pests in the continental United States. U.S. Dep. Agric., Tech. Bull. 1139, 151 pp.

Clausen, C.P. 1978. Introduced parasites and predators of arthropod pests and weeds: A world review. U.S. Dep. Agric., Agric. Res. Serv., Washington, D.C.

Clausen, C.P., S.W. Clancy, and Q.C. Chock. 1965. Biological control of the oriental fruit fly (*Dacus dorsalis* Hendel) and other fruit flies in Hawaii. U.S. Dep. Agric, Tech. Bull. 1322, 102 pp.

Coker, R.C. 1976. Economic impact of the boll weevil. *In* Boll Weevil Suppression, Management, and Elimination Technology. U.S. Dep. Agric., Agric. Res. Serv., ARS–S–71, pp. 3–4.

Coulson, J.R., W. Klassen, R. James Cook, and others. 1979. Notes on biological control of pests in China. U.S. Dep. Agric. 266 pp.

Cross, W.H., and R.H. Chestnut. 1971. Arthropod parasites of the boll weevil, *Anthonomus grandis*: 1. An annotated list. Ann. Entomol. Soc. Am. 64:516–527.

Cross, W.H., and H.C. Mitchell. 1969. Distribution and importance of *Heterolaccus grandis* as parasite of the boll weevil. Ann. Entomol. Soc. Am. 62:235–236.

Danks, H.V., R.L. Rabb, and P.S. Southern. 1979. Biology of insect parasites of *Heliothis* larvae in North Carolina. J. Ga. Entomol. Soc. 14:36–64.

Davich, T.B. 1976. Boll weevil sterility. *In* Boll Weevil Suppression, Management, and Elimination Technology. U.S. Dep. Agric., Agric. Res. Serv., ARS–S–71, pp. 53–58.

DeBach, P. (ed.). 1964. Biological control of insect pests and weeds, 844 pp., illus. Reinhold Publishing, New York.

Dempster, J.P. 1983. The natural control of butterflies and moths. Biol. Rev. 58:461–481.

Dysart, R.J. 1973. The use of *Trichogramma* in the USSR. Proc. Tall Timbers Conf. Ecol. Anim. Control Habitat Manage. 1972:165–173.

Fiske, W.F. 1910. Superparasitism: An important factor in natural control of insects. J. Econ. Entomol. 3:88–97.

Fuester, R.W., J.J. Drea, F. Gruber, and others. 1983. Larval parasites and other natural enemies of *Lymantria dispar* (Lepidoptera: Lymantriidae) in Burgenland, Austria, and Wurtzburg, Germany. Environ. Entomol. 12:724–737.

Fye, R.E., and W.C. McAda. 1972. Laboratory studies of the development, longevity, and fecundity of six species of lepidopterous pests on cotton in Arizona. U.S. Dep. Agric., Tech. Bull. 1454, 72 pp.

Gargiullo, P.M., and C. Wayne Berisford. 1981. Effects of host density and bark thickness on the densities of parasites of the southern pine beetle (*Dendroctonus frontalis* (Zimmerman)). Environ. Entomol. 10:392–399.

Godwin, P.A., and T.M. Odell. 1981. Intensive laboratory and field evaluations of individual species. *In* The Gypsy Moth: Research Toward Integrated Pest Management. U.S. Dep. Agric., Tech. Bull. 1584, 757 pp.

Godwin, P.A., and T.M. Odell. 1984. Laboratory study of competition between *Blepharipa pratensis* and *Parasetigena silvestris* (Diptera-Tachinidae) in *Lymantria dispar* (Lepidoptera: Lymantriidae). Environ. Entomol. 13:1059–1063.

Goodenough, J.L. 1986. Efficacy of entomophagous arthropods. *In* S.J. Johnson, E.G. King, and J.R. Bradley, Jr. (eds.), Theory and Tactics of *Heliothis* Population Management. pp. 79–91. South. Coop. Ser. Bull. 316.

Graham, O.H. 1985. Symposium on eradication of the screwworm from the United States and Mexico. Entomol. Soc. Am. Misc. Publ. 62, 68 pp.

Greany, P.D., T.R. Ashley, R.M. Baranowski, and D.L. Chambers. 1976. Rearing and life history studies on *Biosteres* (*Opius*) *longicaudatus* (Hym.: Braconidae). Entomophaga 21:207–215.

Greany, P.D., J.H. Tumlinson, D.L. Chambers, and G.M. Bousch. 1977. Chemically mediated host finding by *Biosteres* (*Opius*) *longicaudatus*, a parasitoid of tephritid fruit fly larvae. J. Chem. Ecol. 3:189–195.

Griffin, J.G., P.P. Sikorowski, and O.H. Lindig. 1983. Mass rearing boll weevils. *In* Cotton Insect Management with Special Reference to the Boll Weevil. U.S. Dep. Agric., Agric. Handb. 589, pp. 265–282.

Grimble, D.G. 1975. Dispersal of released *Brachymeria intermedia.* NY State Univ., Coll. Environ. Sci. For., Appl. For. Res. Inst. Note 15.

Gross, Harry R., Jr. 1981. Employment of kairomones in the management of parasitoids. *In* D.A. Nordlund, R.L. Jones, and W.J. Lewis (eds.), Semiochemicals: Their Role in Pest Control. pp. 137–150. John Wiley and Sons, New York, Chichester, Brisbane, Toronto.

Gross, H.R., Jr. 1988. *Archytas marmoratus* (Diptera: Tachinidae): Field survival and performance of mechanically extracted maggots. Environ. Entomol. 17:233–237.

Gross, H.R., Jr., and R. Johnson. 1985. *Archytas marmoratus* (Diptera: Tachinidae): Advances in large scale rearing and associated biological studies. J. Econ. Entomol. 78:1350–1353.

Gross, H.R., Jr., and O.P. Young. 1984. *Archytas marmoratus* (Diptera: Tachinidae): Screened-cage evaluations of selected densities of adults against larval populations of *Heliothis zea* and *Spodoptera frugiperda* (Lepidoptera: Noctuidae) on whorl and tassel stage corn. Environ. Entomol. 13:157–161.

Haramoto, F.H., and H.A. Bess. 1970. Recent studies on the abundance of the oriental and Mediterranean fruit flies and the status of their parasites. Proc. Hawaii. Entomol. Soc. 20:551–566.

Hartstack, A.W., J.D. Lopez, R. A. Muller, and others. 1982. Evidence of long range migration of *Heliothis zea* (Boddie) into Texas and Arkansas. Southwest. Entomol. 7:188–201.

Hassan, S.A. (ed.). 1988. Trichogramma News, No. 4, 32 pp. Federal Biological Research Centre for Agriculture and Forestry Messeweg 11/12 D–3300, Braunschweig.

Hensley, S.D. 1971. Management of sugarcane borer populations in Louisiana, a decade of change. Entomophaga 16:133–146.

Hill, R.E., D.P. Carpino, and Z.B. Mayo. 1978. Insect parasites of the European corn borer, *Ostrinia nubilalis*, in Nebraska from 1948–1976. Environ. Entomol. 7:249–253.

Hopper, K.R., and E.G. King. 1984. Preference of *Microplitis croceipes* (Hymenoptera: Braconidae) for instars and species of *Heliothis* (Lepidoptera: Noctuidae). Environ. Entomol. 13:1145–1150.

Huffaker, C.B. 1977. Augmentation of natural enemies in the People's Republic of China. *In* R.L. Ridgway and S.B. Vinson (eds.), Biological Control by Augmentation of Natural Enemies. pp. 329–339. Plenum Press, New York and London. 480 pp.

Huffaker, C.B., R.L. Rabb, and J.A. Logen. 1977. Some aspects of population dynamics relative to augmentation of natural enemy action. *In* R.L. Ridgway and S.B. Vinson (eds.), Biological Control by Augmentation of Natural Enemies. pp. 3–38. Plenum Press, New York and London. 480 pp.

Hughes, P.S. 1975. The biology of *Archytas marmoratus* (Town.). Ann. Entomol. Soc. Am. 68:759–767.

International Atomic Energy Agency. 1982. Sterile insect technique and radiation in insect control. Proc. Symposium, Neuherberg, June 29 to July 3, 1981, 495 pp.

Jackson, C.G., D.E. Bryan, and R. Patana. 1969. Laboratory studies of *Eucelatoria armigera*, a tachinid parasite of *Heliothis* spp. J. Econ. Entomol. 62:907–910.

Janes, H.A., and E.K. Bynum. 1941. Experiments with *Trichogramma minutum* Riley as a control of the sugarcane borer in Louisiana. U.S. Dep. Agric., Tech. Bull. 743, 43 pp, illus.

Jarvis, J.L., and W.D. Guthrie. 1987. Ecological studies of the European corn borer (Lepidoptera: Pyralidae) in Boone County, Iowa. Environ. Entomol. 16:50–58.

Johnson, W.L., W.H. Cross, W.L. McGovern, and H.C. Mitchell. 1973. Biology of *Heterolaccus grandis* Burks in a laboratory culture and its potential as an introduced parasite of the boll weevil in the United States. Environ. Entomol. 2:112–118.

Johnson, S.J., E.G. King, and J.R. Bradley, Jr. 1986. Theory and tactics of *Heliothis* population management. South. Coop. Ser. Bull. 316, 161 pp.

Johnson, S.J., H.N. Pitre, J.E. Powell, and W.L. Sterling. 1986. Control of *Heliothis* spp. by conservation and importation of natural enemies. *In* S.J. Johnson, E.G. King, and J.R. Bradley, Jr. (eds.), Theory and Tactics of *Heliothis* Population Management. pp. 132–154. South. Coop. Ser. Bull. 316, 161 pp.

Jones, R.L. 1981. Chemistry of semiochemicals involved in parasitoid-host and predator-prey relationships. *In* D.A. Nordlund, R.L. Jones, and W.J. Lewis (eds.), Semiochemicals: Their Role in Pest Control. pp. 239–250. J. Wiley and sons, New York, Chichester, Brisbane, Toronto.

Jones, R.L., W.J. Lewis, M.C. Bowman, and others. 1971. Host seeking stimulant for parasites of corn earworm: Isolation, identification, and synthesis. Science. 173:842–843.

King, E.G., G.G. Hartley, D.F. Martin, and others. 1979. Production of the tachinid *Lixophaga diatraeae on* its natural host, the sugarcane borer, and on an unnatural host, the greater wax moth. USDA, SEA Adv. Agric. Technol. South. Ser. No. 3, 16 pp.

King, E.G., and R.D. Jackson (eds.). 1989. Proceedings of the workshop on biological control of *Heliothis*: Increasing the effectivenss of natural enemies. New Delhi, India, Nov. 11–15, 1985. U.S. Dep. Agric., Far Eastern Regional Research Office, New Delhi, India, 550 pp.

King, E.G., and N.C. Leppla (eds.). 1984. Advances and challenges in insect rearing. U.S. Dep. Agric., Agric. Res. Serv. (Southern Region), New Orleans, LA, 306 pp.

King, E.G., L.R. Miles, and D.F. Martin. 1976. Some effects of superparasitism by *Lixophaga diatraeae* of sugarcane borer in the laboratory. Entomol. Exp. Appl. 20:261–269.

King, E.G., J. Sanford, J.W. Smith, and D.F. Martin. 1981. Augmentative release of *Lixophaga diatraeae* (Dip. Tachinidae) for suppression of early-season sugarcane borer populations in Louisiana. Entomophaga 26:59–69.

Klassen, Waldemar. 1989. Eradication of introduced arthropod pests: Theory and historical practice. Entomol. Soc. Am. Misc. Publ. No. 73.

Klun, J.A., J.R. Plimmer, B.A. Bier-Leonhard, and others. 1979. Trace chemicals: The essence of sexual communication systems in *Heliothis* spp. Science 204:1328.

Knipling, E.F. 1971. Use of population models to appraise the role of larval parasites in suppressing *Heliothis* populations. U.S. Dep. Agric. Tech. Bull. 1434, 36 pp.

Knipling, E.F. 1972. Simulated population models to appraise the potential for suppressing sugarcane borer populations by strategic releases of the parasite, *Lixophaga diatraeae*. Environ. Entomol. 1:1–6.

Knipling, E.F. 1977. The theoretical basis for augmentation of natural enemies. *In* R.L. Ridgway and S.B. Vinson (eds.), Biological Control by Augmentation of Natural Enemies. pp. 79–123. Plenum Press, New York and London. 480 pp.

Knipling, E.F. 1979. The basic principles of insect population suppression and management. U.S. Dep. Agric., Agric. Handb. No. 512, 659 pp.

Knipling, E.F. 1980. Regional management of the fall armyworm, a realistic approach? Fla. Entomol. 63:468–480.

Knipling, E.F. 1982. Present status and future trends of the SIT approach to the control of arthropod pests. *In* Sterile Insect Technique and Radiation in Insect Control: Proc. Symp. Leuherberg, IAEA and FAO, pp. 3–23. June 29–July 3, 1981, 495 pp.

Knipling, E.F. 1983. Analysis of technology available for eradication of the boll weevil. *In* R.L. Ridgway, E.P. Lloyd, and W.H. Cross (eds.), Cotton Insect Management With Special Reference to the Boll Weevil. U.S. Dep. Agric., Agric. Handb. 589, pp. 409–436.

Knipling, E.F., and W. Klassen. 1976. Relative efficiency of various genetic mechanisms for suppression of insect populations. U.S. Dep. Agric. Tech. Bull. 1533, 56 pp.

Knipling, E.F., and J.U. McGuire. 1968. Population models to appraise the limitations and potentialities of *Trichogramma* in managing host insect populations. U.S. Dep. Agric. Tech. Bull. 1387, 44 pp.

Knipling, E.F., and E.A. Stabelbacher. 1983. The rationale for areawide management of *Heliothis* (Lepidoptera: Noctuidae) populations. Bull. Entomol. Soc. Am. 29:29–37.

Koyama, J., T. Tanaka, and K. Tanaka. 1984. Eradication of the oriental fruit fly (Diptera: Tephritidae) from the Okinawa Islands by a male annihilation method. J. Econ. Entomol. 77:468–472.

LaChance, L.E. 1985. Genetic methods for the control of lepidopterous species, status and potential. U.S. Dep. Agric., Agric. Res. Serv. ARS–28, 40 pp.

Laster, M.L. 1972. Interspecific hybridization of "*Heliothis virescens*" and *H. subflexa.* Environ. Entomol. 1:682–687.
Laster, M.L., W.F. Kitten, E.F. Knipling, and others. 1987. Estimates of the overwintered population density and adult survival rates for *Heliothis virescens* (Lepidoptera: Noctuidae) in the Mississippi Delta. Environ. Entomol. 16:1076–1081.

Laster, M.L., D.F. Martin, and D.W. Parvin, Jr. 1976. Potential for suppressing tobacco budworm (Lepidoptera: Noctuidae) by genetic sterilization. Miss. Agric. For. Sta. Tech. Bull. 82, 9 pp.

Lawrence, P.O., R.M. Baranowski, and P.D. Greany. 1976. Effect of host age on development of *Biosteres* (= *Opius*) *longicaudatus*, a parasitoid of the Caribbean fruit fly, *Anastrepha suspensa.* Fla. Entomol. 59:33–9.

Lawrence, P.O., P.D. Greany, J.L. Nation, and R.M. Baranowski. 1978. Oviposition behavior of *Biosteres longicaudatus*, a parasite of the Caribbean fruit fly, *Anastrepha suspensa.* Ann. Entomol. Soc. Am. 71:253–256.

Leonard, D.E. 1981. Bioecology of the gypsy moth. *In* The Gypsy Moth: Research Toward Integrated Pest Management. U.S. Dep. Agric., Tech. Bull. 1584.

Lewis, W.J. 1970a. Life history and anatomy of *Microplitis croceipes* (Hymenoptera: Braconidae), a parasite of *Heliothis* spp. (Lepidoptera: Noctuidae). Ann. Entomol. Soc. Am. 63:67–70.

Lewis, W.J. 1970b. Study of species and instars of larval *Heliothis* parasitized by *Microplitis croceipes.* J. Econ. Entomol. 63:363–365.

Lewis, W.J. 1981. Semiochemicals: Their role with changing approaches to pest control. *In* D.A. Nordlund, R.L. Jones, and W. Joe Lewis (eds.), Semiochemicals: Their Role in Pest Control. pp. 3–10. John Wiley and Sons, New York, Chichester, Brisbane, Toronto.

Lewis, L.C. 1982. Present status of introduced parasitoids of the European corn borer, *Ostrinia nubilalis* (Hübner) in Iowa. Iowa State J. Res. 56:429–436.

Lewis, W.J., and J.R. Brazzel. 1968. A three-year study of parasites of the bollworm and the tobacco budworm in Mississippi. J. Econ. Entomol. 61:673–676.

Lewis, W.J., and R. Jones. 1971. Substance that stimulates host-seeking by *Microplitis croceipes*, a parasite of *Heliothis* species. Ann. Entomol. Soc. Am. 64:471–473.

Lewis, W.J., R.L. Jones, D.A. Nordlund, and A.N. Sparks. 1975. Kairomones and their use for management of entomophagous insects. I. Evaluation for increasing rate of parasitization by *Trichogramma* spp. in the field. J. Chem. Ecol. 1:343–347.

Lewis, W.J., and J.H. Tumlinson. 1988. Host detection by chemically mediated associative learning in a parasitic wasp. Nature 331:257–259.

Lindquist, D.A., and E. Busch-Petersen. 1987. Intern. Atomic Energy Agency, Vienna, Austria. Applied Insect Genetics and IPM. *In* Protection Integree: Quo Vadis. "Parasitis 86." pp. 237–255. Intern. Symp. Fruit Flies/Crete, Sept. 1986.

Lloyd, E.P., G.H. McKibben, J.A. Witz, and others. 1981. Mass trapping for detection, suppression, and integration with other suppression measures against the boll weevil. *In* E.R. Mitchell (ed.), Management of Insect Pests with Semiochemicals, Concepts and practices. pp. 191–203. Plenum Press, New York.

Lloyd, E.P., G.H. McKibben, J.E. Leggett, and A.W. Hartstack. 1983. Pheromones for survey, detection, and control. *In* R.L. Ridgway, E.P. Lloyd, and W.H. Cross (eds.), Cotton Insect Management with Special Reference to the Boll Weevil. U.S. Dep. Agric., Agric. Handb. 589, pp. 179–206.

Lloyd, E.P., F.C. Tingle, J.R. McCoy, and T.B. Davich. 1966. The reproduction-diapause approach to population control of the boll weevil. J. Econ. Entomol. 59:813–816.

Loke, W.H., and T.R. Ashley. 1984. Potential uses of kairomones for behavioral manipulation of *Cotesia marginiventris* (Cresson). J. Chem. Ecol. 10:1377–1384.

Loke, W.H., T.R. Ashley, and R.I. Sailer. 1983. Influence of fall armyworm, *Spodoptera frugiperda* (Lepidoptera: Noctuidae), larvae and corn plant damage on host finding in *Apanteles marginiventris* (Gresson) (Hymenoptera: Braconidae). Environ. Entomol. 12:911–915.

Luginbill, P. 1928. The fall armyworm. U.S. Dep. Agric. Tech. Bull. 34, 92 pp.

McGuire, J.U., T.A. Brindley, and T.A. Bancroft. 1957. The distribution of European corn borer larvae, *Pyrausta nubilalis* (HBN), in field corn. J1 paper No. J–2536, Iowa Agric. Exp. Sta., Ames, IA. pp. 65–78.

McPherson, R.M., and S.D. Hensley. 1976. Potential of *Lixophaga diatraeae* for control of *Diatraea saccharalis* in Louisiana. J. Econ. Entomol. 69:215–218.

Martin, W.R., Jr., D.A. Nordlund, and W.C. Nettles, Jr. 1989. Influence of host development state on host suitability and reproductive biology of the parasitoid, *Eucelatoria bryani*. Entomol. Exp. Appl. 50:141–147.

Maxwell, F.G., and P.R. Jennings. 1980. Breeding plants resistant to insects. 683 pp. John Wiley and Sons, New York.

Miller, C.A. 1963. Parasites and the spruce budworm. *In* R.F. Morris (ed.), The Dynamics of Epidemic Spruce Budworm Populations. pp. 228–244. Entomol. Soc. Can. Mem. 31, 332 pp.

Miller, J.H., and G.F. Crisfield. 1930. The presence in Georgia of *Bracon mellitor* Say, a parasite of the cotton boll weevil. J. Econ. Entomol. 23:607–8.

Mitchell, E.R., and J.R. McLaughlin. 1982. Suppression of mating and oviposition by fall armyworm and mating by corn earworm in corn, using the air permeation technique. J. Econ. Entomol. 75:270–274.

Morrison, G., and D.R. Strong. 1980. Spatial variations in host density and the intensity of parasitism: Some empirical examples. Environ. Entomol. 9:149–152.

Mueller, T.F., and J.R. Phillips. 1983. Population dynamics of *Heliothis* spp. in spring weed hosts in southeastern Arkansas and stage-specific parasitism. Environ. Entomol. 12:1846–1850.

Namken L.N., M.D. Heilman, J.N. Jenkins, and P.A. Miller. 1983. Plant resistance and modified cotton cultures. *In* R.L. Ridgway, E.P. Lloyd, and W.H. Cross (eds.), Cotton Insect Managment with Special Reference to the Boll Weevil. U.S. Dep. Agric., Agric. Handb. 589, pp. 73–102.

Nettles, W.C. Jr., and M.L. Burks. 1975. A substance from *Heliothis virescens* larvae stimulating larviposition by females of the tachinid, *Archytas marmoratus*. J. Insect Physiol. 21:965–978.

Nettles, W.C., C.M. Wilson, and S.W. Ziser. 1980. A diet and methods for the in vitro rearing of the tachinid, *Eucelatoria* sp. Ann. Entomol. Soc. Am. 73:180–84.

Newell, I.M., and F.H. Haramoto. 1968. Biotic factors influencing populations of *Dacus dorsalis* in Hawaii. Proc. Hawaii. Entomol. Soc. 20:81–139.

Nordlund, D.A., R.L. Jones, and W.J. Lewis (eds.). 1981. Semiochemicals: Their role in pest control. 306 pp. John Wiley and Sons, New York, Chichester, Brisbane, Toronto.

North, D.T. 1975. Inherited sterility in lepidoptera. Ann. Rev. Entomol. 20:167.

Odell, T.M., and P.A. Godwin. 1979. Laboratory techniques for rearing *Blepharipa pratensis*, a tachinid parasite of gypsy moth. Ann. Entomol. Soc. Am. 72:632–635.

Odell, T.M., and P.A. Godwin. 1984. Host selection by *Blepharipa pratensis* (Meigen), a tachinid parasite of the gypsy moth, *Lymantria dispar* L. J. Chem. Ecol. 10:311–320.

Parencia, C.R., and C.B. Cowan, Jr. 1972. Comparative yields of cotton in treated and untreated plots in insect control experiments in central Texas, 1939–70. J. Econ. Entomol. 65:480–481.

Parencia, C.R., T.R. Pfrimmer, and A.R. Hopkins. 1983. Insecticides for control of cotton insects. *In* Cotton Insect Management with Special Reference to the Boll Weevil. U.S. Dep. Agric., Agric. Handb. 589, pp. 237–264.

Parker, H.L. 1931. *Macrocentrus gifuensis* Ashmead, a polyembryonic parasite in the European corn borer. U.S. Dep. Agric. Tech. Bull. 230, 62 pp.

Pemberton, C.E., and H.F. Willard. 1918. Interrelations of fruit fly parasites in Hawaii. J. Agric. Res. 12:285–295.

Pierce, W.D. 1908. Studies of parasites of the boll weevil. U.S. Dep. Agric, Bur. Entomol. Bull. 73:631–663.

Puterka, G.J., J.E. Slosser, and J.R. Price. 1985. Parasites of *Heliothis* spp. (Lepidoptera: Noctuidae) parasitism and seasonal occurrence for host corps in the Texas Rolling Plains. Environ. Entomol. 14:441–446.

Raulston, J.R., W.W. Wolf, P.D. Lingren, and A.N. Sparks. 1982. Migration as a factor in *Heliothis* managment. Proc. Int. Wrksp. on *Heliothis* Management, November 15–20, 1981, pp. 61–73. Pantancheru, A.P., India.

Ridgway, R.L., E.G. King, and J.L. Carrillo. 1977. Augmentation of natural enemies for control of plant pests in the Western Hemisphere. *In* R.L. Ridgway and S.B. Vinson (eds.), Biological Control by Augmentation of Natural Enemies. 480 pp. Plenum Press, New York, London.

Ridgway, R.L., E.P. Lloyd, and W.H. Cross. 1983. Cotton insect management with special reference to the boll weevil. U.S. Dep. Agric., Agric. Handb. 589, 591 pp.

Ridgway, R.L., and S.B. Vinson (eds.). 1977. Biological control by augmentation of natural enemies. 480 pp. Plenum Press, New York.

Roth, J.P., E.G. King, and S.D. Hensley. 1982. Plant, host, and parasite interaction in the host selection sequence of the tachinid, *Lixophaga diatraeae* (Diptera: Tachinidae). Environ. Entomol. 11:273–277.

Roth, J.P., E.G. King, and A.C. Thompson. 1978. Host location behavior by the tachinid *Lixophaga diatraeae*. Environ. Entomol. 7:794–98.

Rummel, D.R. 1976. Reproduction-diapause boll weevil control. *In* Boll Weevil Suppression, Management and Elimination Technology. U.S. Dep. Agric., Agric. Res. Serv., ARS–S–71, pp. 28–30.

Sailer, R.I. 1972. A look at USDA's biological control of insect pests: 1888 to present. Agric. Sci. Rev. First. Quarter. 1972:15–27.

Semevsky, F.N. 1973. Studies of the dynamics of the gypsy moth, *Porthetria dispar* L. at low population density levels. Entomol. Rev. (English translantion Entomol. Oborzr) 52:25–29.

Shyu, R.C. 1981. Biological studies of *Microncentrus grandii* (Hymenoptera: Braconidae) and its reponse to the kairomones associated with host, *Ostrinia nubilalis* (Lepidoptera: Pyralidae). Thesis submitted to the faculty of the graduate school of the University of Minnesota.

Snow, J.W., and W.W. Copeland. 1969. Fall armyworm: Use of virgin female traps to detect males and to determine seasonal distriubtion. U.S. Dep. Agric. Res. Rept. No. 110, 9 pp.

Snow, J.W., and W.W. Copeland. 1971. Distribution and abundance of the corn earworm in the United States. U.S. Dep. Agric., Agric. Res. Ser., Plant Protection Division Coop. Econ. Insect Report 21:71–77.

Stadelbacher, E.A. 1982. An overview and simulation of tactics for management of *Heliothis* spp. on early season wild host plants. *In* Proceedings of the Beltwide Cotton Production Conferences. pp. 209–212. Las Vegas, NV.

Stadelbacher, E.A., J.E. Powell, and E.G. King. 1984. Parasitism of *Heliothis* zea and H. virescens (Lepidoptera: Noctuidae) larvae in wild and cultivated host plants in the Delta of Mississippi. Environ. Entomol. 13:1167–1172.

Steiner, L.F., W.G. Hart, E.J. Harris, and others. 1970. Eradication of the oriental fruit fly from the Mariana Islands by the methods of male annihilation and sterile insect release. J. Econ. Entomol. 63:131–135.

Steiner, L.F., W.C. Mitchell, E.J. Harris, and others. 1965. Oriental fruit fly eradication by male annihilation. J. Econ. Entomol. 58:961–964.

Steiner, L.F., G.G. Rohwer, E.L. Ayers, and L.D. Christenson. 1961. The role of attractants in recent Mediterranean fruit fly eradication program in Florida. J. Econ. Entomol. 54:30–35.

Stevens, L.M., A.L. Steinhauer, and J.R. Coulson. 1975. Suppression of Mexican bean beetle on soybeans with annual inoculative release of *Pediobius foveolatus*. Environ. Entomol. 4:947–957.

Summers, T.E., E.G. Kind, D.F. Martin, and R.D. Jackson. 1976. Biological control of *Diatraea saccharalis* (Lepidoptera: Pyralidae) in Florida by periodic releases of *Lixophaga diatraeae* (Dip: Tachinidae). Entomophaga 21:359–66.

Summy, K.R., W.G. Hart, M.R. Davis, and others. 1988. Regionwide management of boll weevil in southern Texas. *In* Proceedings of the Beltwide Cotton Production Conferences. pp. 240–247. Las Vegas, NV.

Tardrew, C.G. 1951. The biological control of the karoo caterpillar. Farming S.A. 26:167–168.

Toran, F.O., and Novaretti, W.R. 1978. Management of populations of *Diatraea saccharalis* Fabr. (Lepidoptera: Crambidae) in sugarcane fields of São Paulo, Brazil. Proc. Int. Soc. Sugarcane Technol. 671–687.

Torgersen, T.R., R.W. Campbell, N. Srivastava, and R.C. Beckwith. 1984. Role of parasites in population dynamics of the western spruce budworm (Lepidoptera: Tortricidae) in the northwest *Choristoneura occidentalis* (Freeman). Environ. Entomol. 13:568–573.

Tumlinson, J.H., R.C. Gueldner, D.D. Hardee, and others. 1971. Identification and synthesis of the four compounds comprising the boll weevil sex attractant. J. Org. Chem. 36:2616–2621.

Tumlinson, J.H., P.E.A. Teal, E.R. Mitchell, and R.R. Heath. 1984. Sex pheromones of economically important North American spodoptera species. Proc. Int. Conf. on Integrated Plant Protection. Budapest, Hungary.

U.S. Department of Agriculture. 1976a. Report of Agricultural Research Service's National *Heliothis* Planning Conference. New Orleans, LA, January 19–21, 36 pp.

U.S. Department of Agriculture. 1976b. Boll weevil suppression, management, and elimination technology: Proceedings of a Conference, Memphis, TN, 1974. U.S. Dep. Agric., Agric. Res. Serv., ARS–S–71, 172 pp.

U.S. Department of Agriculture. 1981. The gypsy moth: Research toward integrated pest management. U.S. Dep. Agric. Tech. Bull. 1584, 757 pp.

Vanden Bosch, R., and K.S. Hagen. 1966. Predaceous and parasitic arthropods in California cotton fields. Calif. Agric. Exp. Stn. Bull. 820, Univ. of Calif.-Berkeley. 32 pp.

Vanden Bosch, R., and F.H. Haramoto. 1953. Competition among parasites of the oriental fruit fly. Proc. Hawaii. Entomol. Soc. 15:201–206.

Valentine, H.T. 1983. Budbreak and leaf growth functions for modeling herbivory in some gypsy moth hosts. Forest Service. 39:607–617.

Van Lenteren, J.E. 1981. Host discrimination by parasitoids. *In* D.A. Nordlund, R.L. Jones and W.J. Lewis (eds.), Semiochemicals: Their Role in Pest Control. pp. 153–179. John Wiley and Sons, New York, Chichester, Brisbane, Toronto.

Vargas, R.I., E.J. Harris, and T. Nishida. 1983. Distribution and seasonal occurrence of *Ceratitis capitata* (Wied.) (Diptera: Tephritidae) on the island of Kauai in the Hawaiian Islands. Environ. Entomol. 12:303–310.

Vargas, R.I., D. Miyashita, and T. Nishida. 1984. Life history and demographic parameters of three laboratory-reared tephritids: (Diptera: Tephritidae). Ann. Entomol. Soc. Am. 77:651–656.

Vickery, R.A. 1929. Studies on the fall armyworm in the gulf coast district of Texas. U.S. Dep. Agric. Bull. 138, 63 pp.

Vinson, S.B. 1977. Behavioral chemicals in the augmentation of natural enemies. *In* R.L. Ridgway and S.B. Vinson (eds.), Biological Control by Augmentation of Natural Enemies. pp. 237–279. Plenum Press, NY.

Vinson, S.B. 1981. Habitat location. *In* D.A. Nordlund, R.L. Jones, and W. J. Lewis (eds.), Semiochemicals: Their Role in Pest Control. pp. 51–77. John Wiley and Sons, New York, Chichester, Brisbane, Toronto.

Waage, J., and D. Greathead. 1986. Insect parasitoids. 389 pp. Academic Press, London, Orlando, San Diego, New York, Austin, Boston, Toronto, Sydney, Tokyo.

Weseloh, R.M. 1981. Host location by parasitoids. *In* D.A. Nordlund, R.L. Jones, and W.J. Lewis (eds.), Semiochemicals: Their Role in Pest Control. pp. 75–95. John Wiley and Sons, New York, Chichester, Brisbane, Toronto.

Whitcomb, W.H., and K. Bell. 1964. Predaceous insects, spiders, and mites of Arkanasas cotton fields. Ark. Agric. Exp. Sta. Bull. 69, Univ. Arkansas, Fayetteville, AR, 84 pp.

Winnie, W.V., and H.C. Chiang. 1982. Seasonal history of *Microcentrus grandii* (Hym: Braconidae) and *Eriborus tenebrans* (Hym: Ichnumonidae) two parasitoids of the European corn borer, *Ostrinia nubilalis* (Lep: Pyralidaea). Entomophaga 27:69–74.

Wong, T[im] T.Y., N. Mochizuki, and J.I. Nishimoto. 1984. Seasonal abundance of parasitoids of the Mediterranean and oriental fruit flies (Diptera: Tephritidae) in the Kula area of Maui, Hawaii. Environ. Entomol. 13:140–145.

Wong, T[im] T.Y., and M.M. Ramadan. 1987. Parasitization of the Mediterranean and oriental fruit flies (Diptera: Tephritidae) in the Kula area of Maui, Hawaii. J. Econ. Entomol. 80:77–80.

Wong, Tim T.Y., M.M. Ramadan, D.O. McInnis, and others. 1991. Augmentative releases of *Diachasmimorpha tryoni* (Hymenoptera: Braconidae) to suppress a Mediterranean fruit fly (Diptera: Tephritidae) population in Kula, Maui, Hawaii. Biol. Control 1:2-7.

Wright, J.E., and E.J. Villavaso. 1983. Boll weevil sterility. *In* R.L. Ridgway, E.P. Lloyd, and W.H. Cross (eds.), Cotton Insect Management with Special Reference to the Boll Weevil. U.S. Dep. Agric., Agric. Handb. 589, pp. 153–178.

Zecevic, D. 1958. Daily food consumption of gypsy moth caterpillars on oak trees and on *Pyracantha coccinea*. Zastitabija No. 50, pp. 23–33.

Ziser, S.W., J.A. Wojtowicz, and W.C. Nettles, Jr. 1977. The effect of the number of maggots per host on length of development, pupal, weight, and adult emergence of *Eucelatoria* sp. Ann. Entomol. Soc. Am. 70:733–6.

Appendix

This appendix deals with several important aspects of parasitism that were not discussed adequately in the main body of the report. They will be discussed under separate headings.

Analysis of Some Aspects of the Modeling Procedures

One of the primary purposes of this appendix is to describe in greater detail how the parasite-host population models presented in chapters 4–10 were developed and consider in greater depth the rationales for the theories and parameter values on which the models are based. Also, it seems desirable to critically analyze the biological principles considered in modifying the parameter values used in calculating the influence of the parasite releases on rates of parasitism and on the dynamics of the host populations. The general procedures employed in the current investigation may prove useful for similar investigations on many other parasite-host species.

The degree of confidence that can be placed in the conclusions drawn from this investigation depends on the validity of the theories and assumptions on which the simulation models are based. In the conduct of the investigation, it was necessary to rely on estimates of uncertain accuracy and on assumptions that cannot be supported by definite data. It can be shown, however, that unless certain biological actions take place and unless realistic values are assigned to key parameters, the simulation models will show highly unrealistic and chaotic coexistence patterns between host and associated parasite populations. Without the aid of appropriate models, it is easy to be misled or at least become confused over the way parasitism functions in natural populations. Parasitism data obtained in agricultural areas are often based on limited host collections in habitats where the use of pesticides and/or the cultural practices have seriously distorted normal parasite-host relationships. Data obtained under such conditions may indicate haphazard numerical relationships between hosts and their associated parasite populations and also little rhyme or reason to natural parasitism processes. In fact, however, the numerical relationships between hosts and associated host-dependent parasite populations are regulated by biological actions inherent to nature's balancing scheme and by the biological and behavioral characteristics that have evolved over time between parasites and their hosts. The very existence of so many successful parasites is solid evidence that this is true.

Model A will be used for this analysis of the procedures used to develop models depicting the estimated natural coexistence patterns of parasite and host populations. While the demographics of most of the pest species is not too well known, available general information allows reasonably realistic estimates to be made of the numbers of adult and immature stages that make up typical natural host populations during their annual or periodic cycles. In contrast, virtually no information is available on the actual and relative numbers of even the best known parasite species that normally coexist with their host populations. Indirect procedures to be described were used to make such estimates.

The hypothetical pest host is assumed to produce four generations each year. The adult population during generation 1 averages 100 males and 100 females per acre. The hypothetical parasite attacks the large host larvae, so an estimate must be made for the average number of large larvae produced by the host females each generation. An estimate must also be made for the survival rate of the large unparasitized larvae to the adult host stage. The female host is estimated to produce an average of 50 large larvae during the first generation. The survival rate of the unparasitized host is estimated to be 25 percent. Therefore, the pest population will increase by 5-fold. As the population increases, density-dependent suppression forces will intensify. As a result, fewer larvae will be produced per female and the survival rate will decrease. These influences are reflected in the model. The population will increase

Model A
Basic model depicting the assumed normal seasonal trend of coexisting host and parasite populations and the rate of parasitism for each generation

Parameter (1 acre)	Generation			
	1	2	3	4
Female hosts	100	500	1,600	2,880
Large larvae per female	50	40	30	22.5
Total large larvae	5,000	20,000	48,000	64,800
Female parasites	25	125	400	720
Ratio, parasites to hosts	1:4	1:4	1:4	1:4
Host larvae parasitized per female	40	32	24	18
Total host larvae parasitized	1,000	4,000	9,600	12,960
Percent parasitism	20	20	20	20
Larvae not parasitized	4,000	16,000	38,400	51,840
Survival rates to adults, percent	25	20	15	11
Adult hosts, next generation	1,000	3,200	5,760	5,702
Adult parasites, next generation	250	800	1,440	1,426
Increase rate of host population	5.0	3.2	1.8	1.0
Increase rate of parasite population	5.0	3.2	1.8	1.0

by 3.2-fold during generation 2 and by 1.8-fold during generation 3; then it will stabilize. The seasonal growth rate will be about 30-fold. If 3.3 percent of the adult population survives the winter, the adult female population at the start of the next year will again be about 100 females per acre. While the population trend depicted is considered representative of the population trends of many insect pests, each species has its own number of cycles, typical growth rate, and characteristic density range.

It is well known, however, that insect populations vary greatly from generation to generation and from year to year. The population of a species having the normal density described may on occasion average as few as 10 females per acre or as many as 500 per acre during the first generation. Abnormally low densities, which occur from time to time, are probably not a critical matter for most insect pests, however. When conditions for reproduction return to normal, the increase rate may be as high as 10-fold for several cycles. At this rate, the species would regain its normal low density after one generation and increase well within its normal range after two generations. In all probability it would regain its characteristic range by the next year. Indeed, a population that for any reason has declined to only 1 percent of its normal low level could, under favorable conditions, regain its normal low level after two generations and near its normal high level by season's end.

Of special importance would be information on the coexistence pattern of a host-dependent parasite species under the same variable conditions. Unfortunately, we do not know how parasite populations respond to abnormally low host populations. Simple rationalization suggests, however, that their declines and recuperative powers are similar to those of their hosts. Theoretical models will not be presented, but they indicate that if a host population is 90 percent lower than its normal low level for two generations and thereby causes as much as a 50-percent reduction in the average number of hosts parasitized per female parasite, the rate of parasitism would decline sharply and then, after a long delay, reach its normal level.

The hypothetical parasite used as a model is assumed to be a solitary species that parasitizes the large host larvae. Its life cycle is synchronized with that of its host, and its biology may be reasonably well known. As indicated, it parasitizes about 20 percent of the host larvae. To undertake an appraisal of the way it performs as a natural biological agent and to consider its potential for augmentation purposes, we need more information, such as (1) the approximate number of adults that normally coexist with the host population, (2) a close estimate of the inherent host-finding ability of the female during its lifetime, and (3) the normal rate of survival of the immature parasites to the adult parasite stage. Until such information correlated with information on the hosts is obtained, gaining a clear understanding of the way parasitism functions in natural populations will be difficult.

In the early stages of this investigation, little progress was made in the efforts to develop models that depict approximate actual and relative numbers of parasites that coexist with their host populations—until it was hypothesized that for solitary species, the survival rates of the immature parasites and unparasitized hosts to the respective adult stages tend to be essentially the same. The rationale for this hypothesis has been discussed previously. The hypothesis made it possible to estimate realistic values for all of the key parameters. Like a jigsaw puzzle, the various pieces seemed to fit in the right places and clearly revealed the parasitization picture as a whole. Because of the equal survival hypothesis, information available on the normal rates of parasitism caused by a species can be used to estimate the normal ratios of adult parasites to adult hosts within a rather narrow range of error. In model A, the rate of larval parasitism is assumed to average 20 percent. Therefore, the ratio of adult parasites to adult hosts should average about 20:80 or 1:4. Then, since the host population averages 100 females per acre during generation 1, the parasite population must average about 25 females per acre. Also, since the host population produces 5,000 large larvae per acre and since 20 percent, or 1,000, are normally parasitized, each of the 25 females must parasitize an average of about 40 host larvae during its lifetime. The probable error in the number of hosts parasitized as estimated by this procedure is not likely to be very large. A realistic value for the average number of hosts parasitized by a given parasite species then makes it possible to make a good estimate of the proportion of a known host population that will be parasitized by a given number of parasites. Making such an estimate for the various parasite species discussed in this report was one of the first objectives of the investigation.

Many exploratory models were developed for a number of parasite-host associations that vary greatly in biology and behavior. The purpose was to refine the estimates for the normal parasite-to-host ratios in natural populations; the approximate actual and relative numbers of hosts produced by the host females and of the hosts parasitized by the female parasites; and the actual and relative survival rates of the immature parasites and unparasitized hosts to the respective adult stages. It soon became apparent that minor differences in relative values for these parameters would have a major influence on rates of parasitism and on the host and parasite population trends when projected through three or four cycles. The development of realistic models eventually lead to the following conclusions:

1. The ratio of adult parasites to adult hosts in coexisting populations largely determines the rate of parasitism that will occur, irrespective of the host density. This conclusion has very great practical implications for the augmentation technique.

2. For parasite-host associations whose parasite members are solitary species, the parasite-to-host ratio will normally be directly related to the proportion of the hosts parasitized, because the relative survival rates of the immature parasites and unparasitized hosts to the respective adult stages tend to be about the same at all host densities.

3. Host density has little or no influence on the ratio of adult parasites to adult hosts in coexisting populations. It is environmental factors that largely determine this ratio.

4. The rate of parasitism caused by a parasite species is not indicative of the inherent host-finding ability of the females. The rate merely reflects the relative numbers of adult parasites and adult hosts that normally coexist at their characteristic densities.

5. The actual average number of immature hosts that are of the stage preferred for parasitization and that are produced per female host during its lifetime and the actual average number of hosts parasitized per coexisting female parasite tend to decline as the two populations increase in numbers (a negative correlation between host density and parasite efficiency). However, the relative numbers of hosts produced and hosts parasitized remain reasonably constant at all population densities of the two organisms. The average number of hosts produced per female host generally exceeds the average number parasitized per female parasite.

6. The fecundities of different parasite species reflect what the probabilities are that the immature progeny will survive to the adult parasite stage. The higher the fecundity, the lower the

probability of survival and vice versa. The actual number of hosts that must be parasitized per female depends on the chances of survival of the immature parasite stages in the hosts to the adult parasite stage. In the final analysis, different species that parasitize the same host will be equally successful in terms of the number of adult progeny produced per female parent.

7. The highly developed ability of parasites to find hosts by responding to chemical cues produced by the hosts has virtually eliminated host density as a factor determining the reproductive success of a parasite species. The individual females are able to parasitize their normal lifetime quota of hosts over a wide range of host densities. This ability and the high likelihood that we will eventually be able to achieve and maintain abnormally high parasite-to-host ratios are the real keys to the potential of the parasite augmentation technique as a practical means of regulating pest-host populations.

Every parasite-host association has its own characteristics, but I found that it is not possible to develop population models that reasonably reflect normal population trends and rates of parasitism unless the principles of natural parasitism described are recognized and realistic actual and relative values are assigned to the key parameters. The various factors involved tend to operate in unison under nature's laws to ensure that the relative growth rates of host and associated host-dependent parasite populations will, on average, remain essentially equal, even though actual growth rates vary by manifold. When temporary imbalances occur, which are likely to be frequent, natural regulating factors will restore the normal balance.

The validity of some of the conclusions may be questioned, particularly the conclusion that increases in host density do not increase the efficiency of individual parasites and the rate of parasitism. Many biologists have long assumed that the host-finding efficiency of parasites will increase as the host density increases. That this assumption is invalid can be shown by model B. All of the parameter values for model B are essentially the same as those for the basic model A, except the assumption that the individual female parasites will parasitize 5 percent more hosts for each 100-percent increment of increase in the larval host density. The rate of parasitism during generation 1 would be the same as for the normal population. However, since the larval population during generation 2 has increased by 4-fold, the average number of hosts parasitized

Model B

Parasitism and host and parasite population trends based on the assumption that the female parasite will find 5 percent more hosts for each 100-percent increment of increase in the larval host population (see model A for the estimated normal parasitism and population trends)

Parameter (1 acre)	Generation			
	1	2	3	4
Female hosts	100	500	1,440	1,607
Large host larvae per female	50	40	31	30
Total host larvae	5,000	20,000	44,640	48,210
Female parasites	25	125	560	1,964
Ratio, parasites to hosts	1:4	1:4	1:2.6	1.2:1
Host larvae parasitized per female	44	[1]53	[1]64	[1]66
Total parasite-host encounters	1,100	6,625	35,840	129,624
Ratio, parasite-host encounters to larvae present[2]	.22:1	.33:1	.8:1	2.7:1
Percent parasitism	20	28	55	92
Host larvae parasitized	1,000	5,600	24,552	44,353
Host larvae not parasitized	4,000	14,400	20,088	3,857
Survival to adults, percent	25	20	16	15
Adult hosts next generation	1,000	2,880	3,214	579
Adult parasites, next generation	250	1,120	3,928	6,653
Increase rate of host population	5.0	2.9	1.1	.18
Increase rate of parasite population	5.0	4.5	3.5	1.7

[1]The larval population has increased 4-fold. Thus, the average number of hosts parasitized per female is assumed to increase by 20 percent in generation 2. The larval population in generation 3 has increased by about 9-fold. Therefore, the average number of hosts parasitized per female would increase by 45 percent. In generation 4 the average would increase by about 50 percent.

[2]See table 1 to calculate the rate of parasitism by using the ratio of parasite-host encounters to hosts present.

would be 20 percent higher. The larval population theoretically has increased by about 9-fold by generation 3. Therefore, the average number of hosts parasitized would increase by about 45 percent. These assumptions would mean that the rate of parasitism will be 28 per cent during generation 2, about 55 percent during generation 3, and as high as about 92 percent during generation 4. These are obviously unrealistic rates of parasitism for the kind of parasites evaluated. Thus, for the kind of parasites evaluated, it seemed apparent that even a weak positive response to host density is not typical of their behavior. Indeed, even if the average number of hosts parasitized per female were assumed to remain constant as the parasite and host densities increased, the ratio of adult parasites to adult hosts would become seriously imbalanced within a few generations and the rate of parasitism would increase to an unrealistically high level. The two populations can remain in balance only if the average number of hosts parasitized per female parasite increases or decreases proportionately to the increase or decrease in the average number of hosts produced by the host females.

The accuracy of the estimated average host-finding ability of the various parasite species will largely determine the degree of confidence that can be placed in the results of the suppression models. To indicate what influence a large error in the estimated values for this key parameter will have on parasitism trends and on the dynamics of the host population, I developed models C and D. Values for the parameters are basically the same as those in model A except that we will assume lower or higher host-finding abilities for the individual parasites. For model C the average number of host larvae parasitized per female in each generation will be only half the number estimated for the normal population. The theoretical result will be that the rate of parasitism progressively declines by about half in each succeeding cycle. The host population will grow above its normal level by season's end. A sharply declining rate of parasitism as the host population grows would not be expected for any parasite species.

In contrast, if the female parasites are assumed to find two times as many hosts as assumed for the normal population, the rate of parasitism will theoretically soar to about 98 percent by season's end. It is certain that a host population would have difficulty surviving if it were subjected to a parasite species having such a high degree of efficiency. The theoretical results of these extremes in assumed efficiency are reasons to have confidence that the efficiency estimate made for the hypothetical parasite population depicted in model A is approximately correct.

The theoretical results discussed are merely indicative of the sensitivity of the values assigned to the key parameters used in such models. If model A is accepted as reasonably representative of the coexistence pattern of a given parasite-host association, it could be used to determine the theoretical influence of minor changes in actual and relative values for all of the different parameters. Only a 10-percent change in the relative values assumed for the average numbers of hosts produced and hosts parasitized in each generation, or a 10-percent higher or lower survival rate, would by season's end cause a major change in the actual and relative numbers of parasites and hosts, and the proportion of the host population parasitized. Biologists interested in the subject may

Model C
Population and parasitism trends based on the assumption that the host-finding ability of the female parasite is only 1/2 the estimated normal efficiency, which is shown in model A

Parameter (1 acre)	Generation			
	1	2	3	4
Female hosts	100	562	2,147	3,689
Large host larvae per female	50	40	27	20
Total host larvae	5,000	22,480	57,969	73,780
Female parasites	25	63	101	79
Ratio, parasites to hosts	1:4	1:9	1:21	1:47
Host larvae parasitized per female	20	16	12	9
Host larvae parasitized	500	1,008	1,212	711
Percent parasitism	10	4.5	2.1	1.0
Larvae not parasitized	4,500	21,472	56,757	73,069
Survival to adults, percent	25	20	13	9
Adult hosts, next generation	1,125	4,294	7,378	6,576
Adult parasites, next generation	125	202	158	64
Increase rate of host population	5.6	3.8	1.7	.9
Increase rate of parasite population	2.5	1.6	.8	.4

Model D
Population and parasitism trends based on the assumption that the host-finding ability of the female parasites is 2 times the estimated normal efficiency, which is shown in model A (also, see model C for comparison)

Parameter[1] (1 acre)	Generation			
	1	2	3	4
Female hosts	100	419	812	460
Large host larvae per female	50	41	35	40
Total host larvae	5,000	17,179	28,420	18,400
Female parasites	25	206	992	2,097
Ratio, parasites to hosts	1:4	1:2	1.2:1	4.5:1
Larvae parasitized per female	80	64	48	36
Total parasite-host encounters	2,000	13,184	47,616	75,492
Ratio, parasite-host encounters to larvae present[2]	.4:1	.8:1	1.7:1	4.1:1
Percent parasitism	33	55	82	98
Larvae parasitized	1,650	9,448	23,304	18,032
Larvae not parasitized	3,350	7,731	5,116	368
Survival to adults	25	21	18	20
Adult hosts, next generation	838	1,624	921	74
Adult parasites, next generation	413	1,984	4,195	3,606
Increase rate of host population	4.2	1.9	.6	.08
Increase rate of parasite population	8.3	4.8	2.1	.9

[1]Arbitrary adjustments are made in the average number of host larvae produced per female and the survival rates to the adult stage because of density changes. The average number of hosts parasitized per female parasite is assumed to be about 2 times the average shown in each generation for the basic model A.
[2]See table 1 to calculate the rate of parasitism by using the ratio of parasite-host encounters to hosts present.

wish to employ similar procedures and introduce any factors they consider necessary for estimating the coexistence pattern of parasites and hosts of interest to them. The use of a computer would greatly facilitate such investigations.

Influence of Parasite Releases

When parasites are released in pest-host ecosystems to greatly increase the ratios of adult parasites to adult hosts as described in chapters 4–10, results that can be expected are high increases in the rates of parasitism and sharp declines in the host populations. Little information is available, however, to indicate what may be expected concerning the exact magnitude and effects of these results. Consequently, as with models of natural populations, they must be deduced on the basis of estimates and assumptions. The theoretical results thus deduced are so striking that their validity will likely be questioned by others just as I also questioned their validity during the course of the investigation. Therefore, it seems desirable to analyze the rationales on which the models are based.

Two important aspects of pest control by releasing parasites in pest-host ecosystems do not involve principles of parasitism. They are the estimates of the sizes of typical pest-host populations in nature and the cost of rearing the associated parasites for releases. These estimates are of great practical significance, however, because the ultimate costs and benefits of releasing parasites to achieve satisfactory control will depend on the number and cost of parasites that must be reared and released.

For most of the parasite-host associations included in this study, it is proposed that enough parasites of solitary species be released to achieve about a 2.5:1 ratio of adult parasites to adult hosts. All of the suppression models are based on the premise that the releases will be made when the adult host populations are at their normal low levels. In some cases it may be necessary or advantageous to reduce the host populations to practical manageable levels by other suppressive measures before releasing the parasites.

While a great deal of thought and study were devoted to estimating the actual numbers of the various pests that exist in typical natural populations, the estimates for at least some of the pests may well be too low. If the number that is normally present per unit area exceeds the estimate by 2-fold, for example, the number of parasites required would have to be increased accordingly. In such event the first-year costs would be much higher than assumed.

The continuing annual costs for maintaining adequate control are not, however, influenced by the size of the normal uncontrolled populations.

Of comparable importance with the estimates of the size of natural pest populations are the estimated eventual costs of rearing the various parasite species. Some of my colleagues have expressed the opinion that my estimates are generally too low. This may well be the case. But I have less concern over possible errors in estimating the costs of rearing parasites than over errors in estimating natural pest population densities. There is reason to be greatly confident that the entomologists, engineers, nutritionists, and technicians engaged in insect rearing technology have the ingenuity to develop efficient rearing methods. The costs of rearing the screwworm fly, several tropical fly species, the boll weevil, and some other insects are probably no higher than one-third the cost estimates made before the mass-rearing methods were perfected.

While the two factors discussed have great practical significance, the role that the parasite augmentation techniques can play in future insect-pest management will almost certainly depend largely on the performance of the parasites that are released. As noted before, there are many situations where pest populations are naturally so low or can be purposely reduced to such a low level that it would be practical to release enough parasites to achieve and maintain indefinitely, if necessary, ratios of parasites to hosts ranging from 10:1 to 100:1. In most cases such ratios would exceed those that are likely to occur in natural populations by 50- to 500-fold or more. We can confidently assume that such grossly distorted ratios will result in very high rates of parasitism and the suppression of host populations to very low levels. But we now have no way of knowing for certain whether this assumption is correct, since such distorted numerical relationships have never been created by artificial means throughout host ecosystems and also since the effects of such relationships have never been observed. The theoretical suppression models indicate that the host populations will decline to near elimination before the natural regulating forces can operate with sufficient intensity to reverse the parasitism trends. The question that must be considered is whether the models are based on sound biological principles and on reasonably accurate estimates. The primary purpose of this portion of the appendix is to address this question.

Pest populations inherently have the potential to minimize the adverse effects of any method of control because the reproductive success of the survivors will increase as the populations decline. But in contrast with control methods that are not target-pest specific, those involving the release of pest-host-specific parasites will permit other natural biological organisms to function in a more normal manner. Also, the pest will be subjected to attack by abnormally high populations of a highly specialized organism that will search for its specific host in its normal habitats. In theory, the parasite populations will be maintained at *very high levels* for prolonged periods of time when their host populations are *very low.* In natural populations when a host population is very low, a host-dependent parasite population will also be very low and no increase in parasitism will occur. It is difficult to envision, however, how a pest host could possibly cope with the unusual numerical relationship that can be created by artificial means between parasites and hosts. We will analyze the factors that are likely to become involved under the abnormal conditions described and decide whether the estimates and assumptions made in calculating the influence of the parasite releases seem biologically sound.

Uncontrolled host populations, as noted, tend to increase progressively from their normal low to their normal high levels during the season. However, as indicated in model A, the reproductive success of the individual females in such populations declines progressively with each cycle because of the increasing effect of density-dependent control factors, or natural balancing mechanisms. While each species develops its own response to changing densities, the description given applies generally to all species.

In contrast, females in host populations under intense suppression will respond in the reverse manner. The high parasitism that is theoretically achieved will cause the total host populations to decline progressively, but the reproductive success of the individual females that survive will increase progressively with each cycle. These fundamental differences in the natural and suppressed host populations are reflected in the models shown in chapters 4–10. As the total population of a host declines, the average number of host larvae produced per surviving female will increase and the survival rate of the unparasitized host larvae will also increase. The degrees of adjustments made for these changes are strictly arbitrary but, I believe, plausible. For most of the insect pests, I estimate that the natural populations increase at rates averaging 2.5- to 3-fold per generation. For the populations that are subjected to parasite releases, I estimate that the rates of increase of the survivors will average about 7-fold. If 3- and 7-fold increase rates are projected over four generations, the relative overall growth rates will be 81-

and 2,401-fold. The difference of about 30-fold leads me to believe that the adjustments made in the reproductive success of the individual females that survive to reproduce in the populations exposed to the augmented parasite populations are unlikely to be too low. While the survivors will make valiant efforts to minimize the influence of the parasite releases, they will, in theory, be confronted with a situation in which the ratio of parasites to hosts will increase progressively, especially if parasites are released in more than one cycle. Also, if the released parasites are host specific, all other biological agents will perform in the normal manner.

The performance of individual female parasites in greatly augmented parasite populations can also be expected to differ significantly from that of female parasites in normal populations. Two important factors will tend to reduce the efficiency of the individual females in the augmented populations. The abnormally high populations created by artificial means will be subject to increased predation. This will mean a shorter average reproductive life and a reduction in the average number of hosts parasitized per female during its lifetime. Also, the high rate of parasitism caused by the population as a whole will greatly increase intraspecific competition. To allow for the effects of increased predation in the released population, the average number of hosts found per female is assumed to be reduced by 20 percent. A further reduction in efficiency is assumed when the parasite population increases. Whether adequate allowances have been made for the negative effects of predation may be open to question. If they are not adequate, more parasites would have to be released. For example, if the efficiency of the females is more likely to be reduced by 30 percent during the generation that releases are made, about 20 percent more parasites would have to be released to compensate for the underestimate.

The negative effect of increased intraspecific competition can be estimated with reasonable assurance of accuracy. As the rate of parasitism increases, the increasing probability of parasites to encounter hosts already parasitized can be calculated by the mathematical formula described in chapter 2. For solitary species, the encounter of hosts already parasitized is assumed without exception to represent a complete loss in efficiency. This, however, is likely to be a conservative assumption for parasite species that can discriminate between parasitized and nonparasitized hosts and reject the former. Discrimination behavior in parasites no doubt evolved for the express purpose of minimizing the adverse effects of intraspecific competition. Therefore, for parasite species that can discriminate effectively, their efficiency may have been underestimated.

Another factor that might have a negative effect on the efficiency of some parasite species is superparasitism, because it can lower the survival rates of the immature parasites. Discrimination behavior would also be an important consideration in the assessment of this factor. It is known that some parasite species do not discriminate and, hence, will readily reparasitize previously parasitized hosts. For some species, published data indicate that superparasitism has little or no adverse effect on parasite survival. But for other species, especially egg-larval species, available data indicate that superparasitism increases the premature mortality of the hosts parasitized and, therefore, the survival rate of the immature parasites also. When such data on the effects of superparasitism are available, appropriate adjustments can be made in the efficiency of the parasite species.

In general, I believe that realistic allowances have been made for the negative effects that a greatly augmented parasite population could have on the efficiency of the individual female parasites. If so, there would seem to be no way that the hosts can avoid the effects of the high parasite-to-host ratios created by artificial means and the high rates of parasitism that will result, as indicated in the simulation suppression models.

The primary purpose of this analysis was to call attention to, and estimate the magnitude of, the probable negative effects of abnormally high parasite populations. There are, however, several factors that have not been considered for the suppression models and that might have significant positive effects. One such factor is the natural populations of the parasite species released. They were ignored when estimates were made of the number of parasites to be released. Another factor is the possibility for hosts superparasitized by some of the solitary species to occasionally produce more than one adult parasite progeny. This possibility was not considered in estimating the number of adult parasite progeny that will be produced from one generation to the next. A factor that may be of most practical significance would be the practice of holding parasites in confinement until they are ready to parasitize the hosts. Some of the parasites, especially the tachinids, have a long preoviposition or prelarviposition period. All of the assumed efficiencies of the parasite species are based on estimates of their efficiencies in natural populations. No doubt in nature the mortality of all of the life stages is high. But the adult mortality of parasites to be used in augmentation programs could be reduced by holding the parasites in confinement until they are ready to parasitize the hosts. Their average efficiency upon release might be essentially double that of the natural populations. Lack of adequate nutrition in natural environments may also limit the efficiency of some species of parasites. Therefore, providing proper nutrition for parasites prior to their release could be another significant positive factor in optimizing the effectiveness of the parasite augmentation technique.

The suppression models shown in chapters 4–10 are based on the assumption that the released parasites are of normal behavior. As noted in chapter 2, insect behaviorists have shown that host finding by parasites is in part a learned behavior. To ensure normal host-seeking behavior of released parasites, it may be necessary to provide the proper host-seeking stimuli prior to their release. Providing such stimuli might even enhance their efficiency.

The suppression models are also based on the assumption that all of the released parasites will reach the target area. This degree of precision in operational programs would be highly unlikely. Therefore, in estimating the requirements and costs, I generally assume that in operational programs, the parasite release rate should be two to three times the number indicated in the models.

Thus, there are both probable negative and positive factors to consider in appraising the validity of the values assigned to the various parameters in the suppression models. Parasite release trials conducted in the proper manner will obviously be necessary before the various questions raised can be answered.

Nothing has been said in the foregoing about the possible negative influence of abnormally low densities on the average host-finding ability of parasites. This subject has, however, been discussed in considerable detail in various other sections of this report. Indirect evidence strongly supports the theory that host densities within the normal range for a host species have little or no influence on the host-finding ability of the coexisting female parasites. However, in estimating the influence of the parasite releases, parasite efficiency is assumed to be considerably reduced when host populations decline below an assumed normal low density. This reduction is assumed even though there is no biological basis for it.

While the foregoing discussion deals primarily with the performance of individual parasites and hosts in coexisting populations, the ratio of parasites to hosts in the total populations is considered the dominant factor determining the ultimate effect of the parasite releases. Therefore, when viewed from a practical standpoint, among the most important contributions of this investigation may be the estimates of the actual numbers of hosts in typical natural populations and the estimated influence of various parasite-to-host ratios. Of comparable practical significance is the assumption that parasite rearing technology can eventually be developed that will make it practical to raise normal parasite-to-host ratios initially by 10-fold or more. Then, if the host populations are reduced to abnormally low levels, it should also be practical to maintain many of those populations at parasite-to-host ratios ranging from 10:1 to 100:1 or higher for as long as necessary. If these assumptions are realistic, we should eventually be able to create conditions never experienced by natural host populations, and which have never been created by artificial means.

There is no question that complex biological actions are constantly at work to ensure a reasonably harmonious coexistence of hosts and the parasites that depend on the hosts for reproduction. However, when normal numerical relationships are altered to the degree proposed, certain of the natural regulating forces will intensify and some totally unexpected biological actions may come into play. An example of such an action is the progressive increase in the parasite-to-host ratio and an increasing rate of parasitism when the initial rate of parasitism exceeds 50 percent. Another is the enhancement of natural parasitism by one species when a different species is released. We will never know, however, whether the results of actual parasite releases will conform with the theoretical results unless appropriate trial programs are undertaken with some of the better known parasite-host associations. As noted before, such programs will be demanding in their execution and costly. But if the results obtained come reasonably close to those depicted by the suppression models shown in chapters 4–10, the parasite augmentation technique should offer almost unlimited possibilities for solving many major insect-pest problems in the future both effectively and economically. This technique is also biologically sound and would be environmentally safe. Developing control techniques having these characteristics has been one of the primary goals of scientists engaged in pest management research during recent decades.

In view of the very promising results of this study, there would seem to be justification for public agencies and the agricultural industries to provide the substantial financial and other resources needed for scientists to greatly expand the efforts to develop the full potential of the augmentation technique for use alone or in conjunction with other methods of control. Integration of the augmentation and autocidal techniques can be expected to result in the mutual potentiation of their actions. Such a result would be an expression of an entirely new principle of insect population suppression. Application of the principle could lead to insect population management and eradication strategies of extraordinary effectiveness and efficiency.

Adverse Effects of Parasite and Host Movement on the Effectiveness of the Parasite Augmentation Technique

This theoretical investigation has shown that the release of adequate numbers of parasites throughout pest-host ecosystems at a strategic time in the seasonal cycles of the hosts offers outstanding promise as a method of rigidly regulating insect-pest populations. It has been emphasized repeatedly, however, that the releases must be made throughout the pest-host ecosystems if satisfactory control is to be achieved. To further emphasize this requirement, hypothetical suppression models will be used to indicate the negative effects of parasite and host movement when the release areas are not well isolated.

Model A, as previously discussed, is considered to depict the normal seasonal trends of a host and an associated host-dependent parasite population. Model E shows the theoretical effects of the release of enough parasites during generation 1 to achieve a 2.5:1 ratio of adult parasites to adult hosts throughout the pest-host ecosystem. It is similar to several other suppression models discussed in chapters 4–10. In theory, for some species the rates of parasitism due to the self-perpetuating populations will reach such high levels that the host population will be virtually eliminated by season's end.

Model F was developed to show the effects of releasing highly mobile parasites to control the host population present in a small unisolated area—an area from which many of the parasite progeny could readily move out. In calculating the results, the assumptions were made that half of each generation of parasite progeny produced in the release area will move out of the area and that the host population will increase by 10 percent during generations 2–4 because of immigrants from areas with higher host populations. The estimate of the rate of parasite movement is based partly on data obtained by Sommers et al. (1976) on the movement of the

Model E
Theoretical influence of parasite releases in a completely isolated host ecosystem (see model A for the estimated natural population and parasitism trends for the hypothetical host and parasite populations)

Parameter[1] (1 acre)	Generation			
	1	2	3	4
Female hosts	100	125	84	26
Large host larvae per female	50	48	52	60
Total host larvae	5,000	6,000	4,368	1,560
Female parasites released, generation 1	250	500	636	542
Ratio, parasites to hosts	2.5:1	4:1	7.6:1	21:1
Host larvae parasitized per female	32	26	22	[1]24
Total parasite-host encounters	8,000	13,000	13,992	13,008
Ratio, parasite-host encounters to larvae present	1.6:1	2.2:1	3.2:1	8:1
Percent parasitism	80	88.3	95.5	99
Larvae parasitized	4,000	5,298	4,171	1,544
Larvae not parasitized	1,000	702	197	16
Survival rates to adults, percent	25	24	26	28
Adult hosts, next generation	250	168	51	5
Adult parasites, next generation	1,000	1,272	1,084	432
Increase rate of host population	1.25	.7	.3	.1
Increase rate of parasite population	2.0	1.3	.85	.4

[1]If the average number of hosts parasitized per female were as low as 12, the ratio of parasite-host encounters to hosts present would still be high, i.e., 4.2:1. This ratio would theoretically result in 98 percent parasitism.

sugarcane borer parasite *Lixophaga diatraeae*, as discussed in some detail in chapter 4. Their data indicate little movement of the released parasites but considerable movement of the progeny produced in each cycle. Model F shows a diminishing rate of parasitism during the season. If the host population is compared with that of an uncontrolled population (model A), the results would seem favorable. But if compared with a host population subjected to parasite releases throughout an isolated ecosystem (model E), the results would be very disappointing. If a release experiment were undertaken to validate model E, the experiment would have to be judged unsuccessful. It should be emphasised that many other parasite and host species are likely to move more extensively than those in the model.

Many trial programs to control insects by such techniques as sterile insect releases and use of traps baited with sex pheromones have resulted in disappointment or failure because of insect-pest movement into or out of the nonisolated experimental areas used. Such

Model F
Negative effects of parasite and host movement on rates of parasitism and on host and parasite population trends when parasites are released in a small nonisolated area (see model E for effects when parasites are released in a completely isolated ecosystem)

Parameter[1] (1 acre)	Generation			
	1	2	3	4
Female hosts emerging in release area	100	125	195	478
Immigrant females	—	12	20	48
Total female hosts	100	137	215	526
Host larvae produced per female	50	48	46	40
Total host larvae	5,000	6,000	9,890	21,040
Female parasites released, generation 1	250	500	525	609
Female parasites emigrating	—	250	262	305
Total female parasites reproducing	250	250	262	305
Ratio, adult parasites to adult hosts	2.5:1	2:1	1.2:1	.6:1
Host larvae parasitized per female	32	32	31	28
Total parasite-host encounters	8,000	8,000	8,122	8,540
Ratio, parasite-host encounters to larvae present[2]	1.6:1	1.3:1	.82:1	.41:1
Percent parasitism	80	73	56	34
Larvae parasitized	4,000	4,380	5,538	7,154
Larvae not parasitized	1,000	1,620	4,352	12,500
Survival to adults, percent	25	24	22	20
Adult hosts emerging, next generation	250	389	957	2,500
Adult parasites emerging, next generation	1,000	1,051	1,218	1,431

[1]Some adjustments are made in the average number of host larvae produced per female, the average number of host larvae parasitized per female parasite, and the survival rates of the immature parasites and unparasitized host larvae because of changing densities.
[2]See table 1 to calculate the rates of parasitism and escape from parasitism by using the ratio of parasite-host encounters to hosts present.

disappointing results can be expected to be magnified if highly mobile parasites are released in small nonisolated areas. If the extent of insect movement cannot be measured, which invariably seems to be the case, there is the tendency to attribute failures or disappointing results to the performance of the suppression techniques rather than faulty methods of testing the technique. While proper evaluation of the parasite augmentation technique will prove to be difficult and costly, the potential of the technique is so great that it justifies the efforts and costs involved.

The science of insect-pest management has made outstanding advances during the past few decades in developing new and promising insect suppression technology. In my view, it has not, however, made much progress in applying the new technology for solving insect pest problems in a more effective and ecologically desirable manner. Much of the current basic technology or that which may be developed in the future will remain dormant indefinitely until research workers fully appreciate the need for appropriate evaluation of new technology involving mobile biological organisms, and until adequate support is provided to test the technology in an appropriate manner.

Future Roles of Classical and Augmentation Biological Control Techniques for Insect-Pest Management

The conclusions reached on the factors that influence insect parasitism, as reviewed in the foregoing, suggest the need to consider the roles of both classical and augmentation biological control techniques for insect-pest management in the future. The conservation of existing biological agents, habitat improvement when practical, and the introduction of new agents especially for alien insect pests are fundamental to sound pest-management practices. Numerous potential pests are controlled effectively by native and introduced biological agents. Introduced biological agents also help prevent many other pest species from causing excessive damage. The pest density that natural biological agents must maintain if they are to provide satisfactory control depends, however, on the nature of the pest and the sensitivity of the crops or animals to damage. Thus, no simple criteria for satisfactory natural biological control can be established. As discussed in the report, the evidence is clear, however, that natural control forces and biological and behavioral characteristics that have evolved tend to maintain the numerical relationship between parasites and hosts within certain limits. This relationship is entirely consistent with nature's laws, but it also virtually precludes the possibility that self-perpetuating biological agents can satisfactorily control many pest species. This conclusion is supported by past experience that most of the major and persistent insect pests have continued to cause unacceptable losses or threats to agriculture for decades despite the introduction of new biological agents and many advances and improvements in other methods of control.

The presumption that the classical biological approach to control can ultimately solve most of the major insect-pest problems if pursued with sufficient intensity cannot be supported either by past experiences or by the fundamental principles governing natural parasitism. However, if administrators, growers, budget officials, environmentalists, and many biological control experts themselves believe that most of our major insect-pest problems can ultimately be met in a satisfactory manner by this very desirable and low-cost method of control, they will be guided by false hopes. They will also have less interest in—and will be unlikely to lend the support needed to adequately develop—other equally acceptable methods of control that may offer greater chances of success.

I have presented an appraisal of the potential of the parasite augmentation technique for rigidly regulating populations of a number of important insect pests. The appraisal indicates that this biological technique offers almost unlimited opportunities for solving many persistent problems caused by insect pests that do not yield adequately to natural enemies. There is little doubt that the barriers that limit the role self-perpetuating populations play as natural control agents can be overcome by releasing key and largely host-specific parasite species into the pest ecosystems in adequate numbers and in the proper manner.

There is every reason to have confidence that truly major advances in insect control by biological means can be achieved in the future if the merits, as well as the limitations, of both the classical and augmentation techniques are fully recognized, and if concerted and coordinated efforts are made to fully exploit both techniques. Prior investigations by biological control experts have shown that there exists a virtually unlimited resource of natural biological agents from which to choose candidate species for both techniques. The characteristics of biological agents that will make them superior for augmentation purposes may, however, differ considerably from those of biological agents most likely to contribute to natural control. Therefore, the objective of exploration programs should be broadened to include the selection and evaluation of biological agents that might be particularly useful for augmentation purposes.

Principles of Insect Predation Versus Insect Parasitism

While little has been said about insect predation in this report, no inference should be drawn that insect predators are less important than insect parasites as natural biological control agents. It is probable that for most insect pests the total predator complex contributes to natural control more than does the total parasite complex.

No effort was made, however, to estimate the potential of predator species for controlling insect pests when released in pest ecosystems. There are no good clues for estimating the actual and relative numbers of a given predator and a given prey species that coexist in natural populations. Most predators are generalists in preying behavior. Therefore, it is difficult to determine the proportion of a pest population destroyed by a given predator. In fact, it is virtually impossible to determine the incidence of predation in natural populations because predators as a rule consume their prey, leaving no evidence of their attack. In contrast, the proportion of a host population parasitized by a given parasite species can be determined by examining host collections. Also, the relative numbers of certain parasite species and their coexisting, associated host populations can be estimated on the basis of the equal survival theory, as previously noted. The relationship between rate of parasitism and the relative numbers of a parasite species and its host (parasite-to-host ratio) can be used to calculate the rates of parasitism that may be anticipated at other parasite-to-host ratios. Such clues are not available for estimating the relative numbers of a predator and its prey in natural populations and, hence, what the numbers mean in terms of the proportion of the prey population controlled by the predator.

Despite the limited information we have or can obtain on the role that predators play as natural biological agents, and the role they might be made to play when used for augmentation purposes, we can reach certain important conclusions by simple rationalizations. There is no question that predators are responsible for high insect-pest mortalities. But in general, they are no doubt also responsible for comparable mortalities of the coexisting insect parasites. To a predator the adults of a parasite species or the host larvae or pupae that contain immature parasites are probably as welcome as a food resource as the adult hosts or host larvae and pupae that escape parasitism. Thus, predators can have a major effect on the *actual* numbers of certain pests and their parasites in natural populations, but they are unlikely to have any influence on the *relative* numbers of the pests and parasites. Therefore, high rates of predation will not alter the very important theory that the ratio of parasites to hosts is the dominant factor that determines the rate of parasitism caused by a parasite species, irrespective of the host density.

In the models depicting the coexistence patterns of the different parasite-host associations, high natural mortalities are assumed for both the host and parasite populations. Except for parasitism, at the rates assumed in the models, the causes of the mortalities are not specified. However, when abiotic conditions are normal and favorable for host and parasite reproduction, most of the natural mortality of both the host and parasite populations is undoubtedly due to predation. That the relative rates of host predation and parasite predation will tend to remain reasonably constant is an assumption that can be supported by modeling procedures. If a hypothetical population model is developed on the assumption that coexisting predators will consistently kill a substantially higher proportion of a parasite population than the host population (for example, 25 percent), the parasite would in time probably disappear. If the proportion is reversed, the parasite population would in time cause near elimination of the host population, a consequence that in turn would cause near elimination of the host-dependent parasite species. Insect-predator and insect-parasite complexes, which usually involve many species, have coexisted in certain pest environments for millions of years. They have evolved abilities that will ensure continued coexistence in a satisfactory manner despite the intense interspecific competition.

While the details of predation and parasitism may differ greatly, the principles involved in the control of prey and hosts are basically the same. The degree of control caused by a given predator will largely depend on the relative numbers of the predator and the prey species in coexisting populations. If a well-adapted predator with a high preference for a certain pest were released in the pest ecosystem in sufficient numbers to greatly exceed the normal ratio, the proportion of the pest population that would be controlled is likely to be similar to the proportion that would be controlled if a parasite species were released instead to achieve a comparable parasite-to-host ratio. The relative merits of the two organisms will depend largely on the relative costs of rearing and releasing them.

The number of predator species that are primarily or totally dependent on a given insect pest for reproduction is smaller than the number of parasite species having this characteristic. This investigation places emphasis on target-pest-specific methods of regulating pest populations. The release of a broad-spectrum predator in a pest ecosystem may cause considerable ecological disruption, although certainly much less than the application of a broad-spectrum insecticide.

In localized habitats, the release of immature predators or adult predators that cannot fly would probably be more advantageous than the release of adult parasites. Such predators can be expected to remain in the release areas. Therefore, the augmentative release of promising predators should provide a much more reliable means of controlling certain pests on a farm-by-farm basis than the release of mobile adult parasites.

Principles of Insect Eradication and Population Management

This investigation was focused on the potential of the parasite augmentation technique for maintaining populations of certain major pests below economically damaging levels. However, because of its characteristics, the technique may also have the potential for eradicating certain pests from prescribed areas, especially when integrated with autocidal techniques. The requirements for eradicating populations and for rigidly regulating populations on an areawide basis are basically similar. But eradication is so demanding that eradicative efforts would not be justified for many pests. Still, if even only a few of the more damaging and persistent species could eventually be eradicated from specified areas in an ecologically acceptable manner, an important contribution will have been made to agriculture and environmental hazards will be significantly alleviated. Thus, eradication of insect pests from prescribed areas should be considered an important aspect of insect-pest management.

Whether or not eradication should be considered one of the important options for dealing with important insect problems is, however, a highly controversial issue among entomologists. The controversy is due to the opinion of many biologists that eradication is not biologically, operationally, and economically feasible for endemic and long-established alien pests. Therefore, it seems desirable to examine the insect eradication issue from the standpoint of the basic principles of suppression that would be involved. The justification for undertaking eradicative or areawide management programs will of course depend on the nature of the pest, the suppression and detection methods that are available, how efficiently the methods can be applied, and the long-range benefits in relation to costs. Most of these variables are subject to change, however, depending on advances that are made in the area of insect-pest management. But the basic principles involved in the eradication or management of populations will not change.

Most entomologists seem to agree that efforts should be made to eradicate new introductions of potentially damaging insect pests if there are reasonable chances of success. In a recent report, Klassen (1989) presented information showing that such eradications have been accomplished on numerous occasions in the past, thereby preventing even more losses to agriculture and obviating the need for using more insecticides than now needed against the pests already present. The major controversy over this issue involves well-established alien pests and endemic pests.

Some biologists have adopted and advanced the theory that elimination of a pest from the fringe areas of its range of distribution may be possible but that some unknown biological factor or factors prevent its elimination from the heart of its endemic areas. A similar theory holds that efforts to eradicate recently established pest populations may prove successful but that such efforts will fail against populations of long-established alien species. I do not know how these theories originated or on what grounds, but they are totally without foundation. They cannot be supported by experience, and they are not consistent with the biological factors that influence the dynamics of insect populations or with the basic principles of pest population suppression.

It is well known that populations of insect pests that are inadvertently introduced, or that spread on their own, into new favorable environments generally reach very high levels and cause serious damage. The populations develop high densities because their rates of increase per cycle are above normal for the species. In time, the populations will decline and then stabilize at a lower level because native biological agents will adapt to the new pests and use them as host or prey resources. A decline and stabilization will also occur if biological agents from the native homes of the pests are introduced and become established. As a rule, however, insect pests tend to remain at higher levels and cause more damage in new environments where they have long established themselves than in their native environments.

These biological phenomena are very much involved with the basic principles of insect population suppression. The higher the characteristic density of a pest in a given ecosystem and the higher its inherent increase rate per cycle, the more its reproductive capability must be inhibited to slow down the rate of growth or cause a decline in the population. Therefore, in actuality, it should be possible to more readily eliminate or regulate populations of endemic species than populations of long-established alien pests. For the same reasons it should be possible to more readily eliminate or manage populations of long-established alien pests than populations of recently established pests in environments of the same size and type. However, the main point I wish to make is that there are no mysterious biological phenomena that will preclude successful eradication or population management programs, whether the pests are endemic, long-established alien, or recently established alien species. Decisions on the feasibility and economic justification for such programs should be based on critical and objective analyses of the nature of the pest, the control technology available, and the costs and benefits compared with those of current control practices.

There is no intent to leave the impression that the eradication or continuous management of insect-pest populations can be easily accomplished. On the contrary, for every species that may be a good candidate for either approach, it will be a major challenge for the research scientists to develop and demonstrate the technology, and an equally formidable challenge for the pest managers to execute programs effectively and efficiently. There are reasons, however, to have confidence that these challenges can be met for many important pests and that the logic of the organized preventive approach to the solution of major pest problems will in time be given more recognition and be more generally accepted and supported by the pest-management community and by growers who experience heavy losses due to certain insect pests.

Many important advances have been made in recent years to develop new and improved insect control methods. But as previously noted relatively little progress has been made in putting the new information into practice. I feel that in large measure this slow progress has been due to misconceptions of the nature discussed and a general lack of understanding of the basic principles of insect population suppression, and what we can and cannot expect from the technology we now have or might expect in the future. Also, the concept of rigidly managing insect-pest populations on an ecosystem basis has received little support by the insect-pest-management community.

A critical analysis of the procedures followed and the results obtained in the execution of the various management and eradication programs for the screwworm (*Cochliomyia hominivorax* (Coquerel)) can give us a broad perspective of the fundamental principles involved in insect population management and eradication programs, irrespective of the suppression methods employed. The screwworm programs and their results are discussed in considerable detail by contributors to the publication edited by Graham (1985). The objectives and results achieved in the various programs will be reviewed briefly, keeping in mind the basic principles of pest population suppression that were involved. If the rate of reproduction of an insect pest or any animal population is inhibited enough by any manner to cause a consistent decline in the population, the population will in time become extinct.

It may be difficult to execute programs for achieving such inhibition, and the decline of the pest population may be a gradual process; but the end result will be inevitable.

The rather small screwworm population in Curaçao and the much larger population in the Southeastern United States were completely isolated. Also, resources were available to subject the total populations to enough sterile flies to cause progressive declines in the native fly populations each generation. Therefore, complete eradication was achieved within a relatively short period. Since there were no sources of screwworms to repopulate the ecosystems by direct flight, the eradication was permanent. The possibility of reintroductions by other means, however, is a continuing threat.

These successes lead the livestock interests in the Southwestern United States to press for an eradication program. A program was proposed, but the researchers involved fully recognized and advised the livestock growers and the Federal and State administrators that permanent eradication of the screwworm from the Southwest would not be biologically feasible if the program were undertaken on the scale under consideration. The screwworm population in the Southwest was only a small part of a much more extensive endemic population that existed throughout Mexico and Central America. The scope of the program as proposed was not large enough to expose a very large proportion of the population in Mexico to sterile fly releases. Indeed, it would not have been possible to release enough sterile flies to expect adequate suppression of the high population that normally existed in the Southwest during the summer months. However, the U.S. fly population during the winter normally declined to relatively low numbers and was restricted to relatively small areas. Therefore, the U.S. population could be attacked at a strategic time and place—conditions that I consider are a fundamental requirement for any program to manage a total pest population. Plans called for the production and release of 100 million flies per week. This release rate was believed to be high enough to result in adequate overflooding of the overwintering population in south Texas. The strategy was to attempt eradication of the U.S. population during the winter and spring months and then establish a permanent sterile fly barrier about 100 miles wide to prevent repopulation by flies from Mexico. No assurance could be given that the program would be successful, because the maximum flight range of the flies was not known. Despite this uncertainty, the representative of the livestock industry virtually demanded that the program be undertaken. The pest had plagued the industry for years, causing high losses; and any effort that had a reasonable chance of success was considered worth the risks. The U.S. Department of Agriculture scientists who developed the technique to be used, including R.C. Bushland, A.W. Lindquist, A.H. Baumhover, and me, were of the same opinion.

The program got under way in Texas in 1962. During the winter of 1962–63, the overwintering population in Texas was virtually reduced to zero. There was optimism that eradication of the U.S. population might be successful and that the success could be maintained by a sterile fly barrier. However, in the early spring of 1963, relatively few but widespread screwworm infestations occurred within the sterile fly release zone, and some infested animals were found as far as about 100 miles north of the sterile fly release area. It was difficult at first to accept the possibility that these infestations were the result of migrating flies from non-fly-release areas in Mexico. Yet, the same pattern occurred the next year. Therefore, a release-and-recapture experiment was conducted with marked sterile flies. The results demonstrated that some sterile flies moved up to 180 miles from the point of release. Other evidence indicated, however, that some native flies probably moved as much as 300 miles. Thus, from a biological standpoint, the program had to be categorized as a population management program. It proved to be remarkably effective, however, despite considerable long-range fly movement. The livestock growers were elated over its success. For the first time in history the screwworm was a pest of little economic significance. I estimate that during the first 5 years or so, the incidence of screwworm infestations in the Southwestern United States was reduced by at least 98 percent. The area of normal summer spread of the pest was also greatly reduced. The savings to the livestock growers were conservatively estimated to exceed the cost by 10-fold or more.

Some entomologists attempted to discredit the program on grounds that the original objective had not been achieved. But they largely ignored the outstanding success of the program when viewed from a screwworm management standpoint. In my opinion, this success was among the most significant achievements of the various programs. The results of the program were the first demonstration that a wide-ranging insect pest could be effectively controlled by a large-scale and well-organized suppression program. The feasibility of managing total populations is not limited to the sterility technique. Any method of control that is applied on a similar scale and that inhibits reproduction to a similar degree would be equally successful from a population management standpoint.

As the program continued, however, there were indications that its effectiveness was gradually diminishing. In general, more infested livestock were reported each year. Then, in 1972, a year exceptionally favorable for screwworm reproduction, the number of infested animals increased to alarming numbers. While the situation improved considerably during the next few years, it was apparent that the effectiveness of the program was declining. This opened the floodgates for skepticism and criticism. Some biologists advanced various theories to account for the reduced effectiveness, even though they may have never seen a screwworm fly or an infested animal and have had no experience with the sterility technique. The most popular theory was that the screwworm had in some manner developed resistance to the sterility technique. Some critics assumed that the program was a complete failure and advocated its discontinuance. I predicted that if this were done, the pest would cause catastrophic losses of livestock and wild animals in many areas of the Southwest and even in the Midwest until livestock producers could again employ necessary personnel and reinstitute former livestock management practices.

The possibility could not be ruled out that some behavioral changes in the native flies and/or the sterile flies were at least partly responsible for the reduced effectiveness. But I rationalized also that much of the problem might be due to a gradually declining ratio of sterile flies to wild flies in the ecosystem, as a result of ecological changes that had taken place. Because the program had been so effective during the early years, most of the livestock producers did not retain enough range riders to monitor the livestock and treat the few infestations that did occur. Livestock management practices that had been followed diligently for years to reduce the number of animals susceptible to screwworm attack during the fly season had since been gradually abandoned. The deer population over a 10-year period had probably increased by severalfold and was distributed over a larger area. The number of livestock had increased also. Furthermore, to economize in operating the program, the flight lanes of the aircraft distributing the flies had been widened. The result was less uniform distribution of the sterile flies. However, the number of flies reared for release remained about the same. I estimate that these negative factors may have increased the probability of animals becoming infested by a factor of 5-fold or more. In effect, the ratio of sterile flies to fertile flies had probably declined accordingly.

It was apparent that to maintain the high degree of control obtained during the earlier years or to achieve the original objective, it would be necessary to expand the program. Fortunately, representatives of the U.S. and Mexican Governments endorsed plans for a greatly expanded program. A fly-rearing facility that had the capacity to produce 500 million flies per week was constructed in Mexico. This facility made it possible for sterile flies to be released in much larger areas in Mexico and the United States, and in larger numbers in critical areas. The new program got under way in 1977, and its benefits became apparent within a year. The number of screwworm cases in the United States was very low in 1977, and only a few cases occurred during the next few years. These straggling cases were not necessarily evidence of survivors of the original population. They were probably due to flies that had traveled long distances from Mexico, or to the movement of infested animals. The number of screwworm cases began declining progressively in northern Mexico during the late seventies and in central Mexico during the early eighties. By 1984 the screwworm had been pushed southward to the narrow Isthmus of Tehuantepec in southern Mexico, where a wide sterile fly barrier could be established with relatively few flies.

Several very important conclusions of fundamental significance to the concepts of insect eradication and insect population management can be drawn from the results of the programs conducted in the Southwestern United States and Mexico. The initial program in Texas soon confirmed that a pest population cannot be eradicated unless the total population within its normal range of movement is subjected to adequate suppression. The program did demonstrate, however, that a very high degree of control of a strong-flying insect pest can be achieved with great benefits to agricultural producers if the suppression program is well organized and undertaken on a large enough scale. Excellent control of the screwworm was achieved despite its considerable movement into the area under management. The success of the expanded program clearly indicated that the gradually diminishing effectiveness of the initial program was largely due to changes in screwworm surveillance and management practices by the livestock growers and not to a failure of the suppression technique. The success invalidated the widely held theory that the native flies had developed resistance to the sterility technique. It also disproved other theories that had been proposed. Finally, it demonstrated that a large and widely distributed pest population can be eradicated by stages if eradication efforts are undertaken in areas larger than the normal dispersal range of the pest.

Although many difficult operational, technical, logistical, and other problems were encountered in the implementation of the various programs, all of the objectives were ultimately achieved. The programs have already saved the livestock industries in the United States and Mexico billions of dollars. The costs of the programs were probably less than one-tenth the financial losses the pest would have caused if not controlled. The success of the several screwworm programs is recognized as an outstanding achievement in solving a major pest problem. But I feel that the basic principles of insect population eradication and population management demonstrated by the programs, if fully recognized by the pest-management community, will in time be of even greater importance to insect-pest management. The same principles should apply for other important species and other suppression techniques with comparable benefits in relation to costs. Insect pests that are suitable candidates for eradication or management on an areawide basis will differ in characteristics; therefore, the best techniques and strategies to be employed against them will also differ. However, the basic principles will not change. The scope of the programs necessary for success will largely depend on the dispersal behavior of the pest and the numbers in which they move into the area needing protection. Obviously, the benefits of eradication or continuous management must exceed the costs. However, a practical matter to keep in mind is that the more important a pest species is in terms of economic losses and the adverse environmental impacts of the control measures now employed for its control, the larger can be the investment in environmentally safe management or eradication programs.

Some of the important pest insects have the ability to disperse in considerable numbers up to several hundred miles during a single cycle and even longer distances during a season. Therefore, to be effective, suppression programs for such species may require the application of control measures on a regional scale. Even if the release of parasites or some other biological agent can be shown to effectively regulate pest populations, a major obstacle to the implementation of suppression programs may be reluctance on the part of scientists and administrators to endorse them merely because of their magnitude. The production of billions of insects within a period of a few months for distribution in areas encompassing several hundred thousand square miles may intuitively be considered out of the question. However, had such a view resulted in rejecting implementation of the screwworm programs, this damaging pest would still be in the United States.

The success in recent years of large-scale programs to suppress populations of the screwworm, medfly, and boll weevil is evidence that the wide scope of a program should not be a matter of great concern in decisions to determine the feasibility of its implementation. The decision to undertake any pest management program by the organized, total-population approach should be based on a comparison of the potential benefits and costs of the approach versus the benefits and costs of the uncoordinated management approach that has been applied on a farm-to-farm basis and relied upon for decades. The relative environmental impacts of the two approaches should also be an important consideration in the decision-making. The discussions presented in chapters 4–10 indicate that the parasite augmentation technique has the potential of broadening the feasibility of managing or eradicating insect pest population in an effective and economical manner by procedures that would be environmentally safe.

Subject Index

Author Index

Biography

Edward F. Knipling is perhaps best known for originating the concept of suppressing insect pest populations by the release of sterile insects into the natural populations. This concept is the basis of the technique developed to eradicate the screwworm from all of the United States and Mexico. The technique has already saved livestock industries in the United States and Mexico billions of dollars in losses to the screwworm. The same concept of releasing sterile insects is used in some countries to eradicate or suppress tropical fruit flies, the most important insect pests of fruit in many areas.

Dr. Knipling was born on March 20, 1909, and was raised on a farm near Port Lavaca, TX. He received a B.S. degree in entomology from Texas A&M University, and an M.S. degree in entomology (1932) and a Ph.D. in entomology and parasitology (1947) from Iowa State University. His long career with the U.S. Department of Agriculture (USDA) began in 1931.

During World War II, Dr. Knipling directed research to develop insecticides and insect repellents for use by the U.S. armed forces to control vectors of important diseases in various war theaters. After the war, Dr. Knipling was assigned to lead the Agricultural Research Service's (ARS') program on control of "insects affecting man and animals." In 1953 he was named director of all of ARS' research programs on insects. During his 17-year tenure as director, he emphasized the need for basic and applied research on biological and other ecologically sound methods of insect pest control.

Dr. Knipling has long advocated that total populations of the more damaging pests be controlled by preventive methods that are target pest specific and environmentally safe. An authority on the dynamics of insect populations, he uses modeling procedures to critically analyze the suppression characteristics of various methods of control, to predict how the methods will influence the dynamics of various pests, and to evaluate how the methods might be used, alone or integrated, for optimum results. He retired in 1973 but has continued to devote much of his time to this work.

Dr. Knipling has received numerous honors and awards for his contributions to entomology and insect pest management. They include the U.S.A. Typhus Commission Medal, U.S.A. Award for Merit, President's Award for Distinguished Federal Civilian Service, National Medal of Science, USDA Distinguished Service Award, and Rockefellower Public Service Award. Texas A&M University and Iowa State University gave him distinguished alumnus awards. He has also received honorary doctor of science degrees from several universities. He is a member of the ARS Science Hall of Fame, National Academy of Sciences, and American Academy of Arts and Sciences.

GPO Order Form

☐ **YES,** please send me ____copies of **Principles of Insect Parasitism Analyzed From New Perspectives**, S/N 001-000-04582-2, at $9.50 each.
(Order processing code: ***6237**)

The total cost of my order is $______. International customers please add 25%. Price includes regular domestic postage and handling and is subject to change.

Company or personal name

Additional address/attention line

Street address

City, state, ZIP Code

Daytime phone including area code

Purchase order number

May we make your name/address available to others? Yes___ No___

To fax your orders, call (202) 512-2250. Or you can charge your order. It's easy. Please choose a method of payment.

☐ Check payable to Superintendent of Documents

☐ GPO Deposit Account ☐☐☐☐☐☐☐-☐

☐ VISA or MasterCard Account

VISA MasterCard

(Credit card expiration date)

Authorizing signature

Mail to Superintendent of Documents, P.O. Box 371954, Pittsburgh, PA 15250-7954

Thank You For Your Order